MALAYSIAN MELIPONICULTURE & BEYOND

Inc. Stingless Bee Conservation

Coordinator: Abu Hassan Jalil
Editor: David W. Roubik

INTERNATIONAL BEE RESEARCH ASSOCIATION

Jointly published by: The International Bee Research Association,
a Company Limited by Guarantee, 1, Agincourt Street, Monmouth, NP25 3DZ (UK) &
Northern Bee Books, Scout Bottom Farm, Mytholmroyd, Hebden Bridge HX7 SJS (UK).

Obtainable from:
www.ibra.org.uk & www.northernbeebooks.co.uk

ISBN: 978-1-913811-06-8

Cover design and artwork by DM Design & Print
IBRA Proof Editor - Stuart A. Roberts

Title: Malaysian Meliponiculture & Beyond

Subtitle: Inc. Stingless Bee Conservation

Synopsis: This is the second instalment to the "Meliponiculture and Beyond" Series. It covers the meliponiculture activities and the variety of produce from stingless beekeepers and conservationists' efforts in Peninsular Malaysia and East Malaysia on Borneo Island. Scrutinizing the regional stingless bee fauna and flora diversity and exploring it with crop pollination potentials. The prospects of highland Stingless Beekeeping is examined with the establishment of a highland station stingless bee repository. It further explores historical data of the 19th century up to the current discoveries on the diverse sugar profile in stingless bee honey. It concludes with some aspirations on the future outlook of the meliponiculture industry of the country.

Acknowledgements

We are much obliged to our mentor Dr. David W. Roubik of Smithsonian Tropical Res. Inst., Panama for editing this book. His comments are mostly preserved in the many footnotes. A compilation of these footnotes would be a valuable booklet in itself.

We are indebted to our advisors, Prof Siti Amrah Sulaiman and Prof Mohd Mansor Ismail, for the forewords they provided for this book.

Special thanks to the technologists and scientists of the Malaysian Genome Institute, who contributed much to the production of this manuscript. Those deserving a mention are Mohd Noor Isa, Mohd Razif Mamat, Dr. Nor Azfa Johari and Dr. Azira Mohamad.

Much appreciation to the contributing scientific authors, namely, Dr. Wan Iryani, Wan Ismail, Dr. Norhasnida Zawawi, Dr. Norela Sulaiman and Dr. Mervynn Koon.

Thanks to contributing authors Rusphin Asyraf Taip of Ranau, Sabah; Muhammad Fahdli Abdul Rahman of Merlimau of Merlimau, Melaka; Hendra Ja of Tg. Malim, Perak; and Jeffry Iskandar Foo of Jelebu, Negeri Sembilan.

Without the contributions of all the above and the hospitality of the many Stingless Beekeepers listed in Chapter 16 who took part in the "The Itama Journey", this book may not have materialized completely.

Contents

Foreword I 10

Foreword II 11

Preface 12

Chapter 1 13

1.0 Pollination Initiatives in Development of Meliponiculture with Cash Crop Agriculture 13

1.1 Abstract: 13

1.2 Introduction 13

1.4 Recommendations 16

1.6 Indications of crop benefits from insect pollination 18

1.7 Does a crop need insect pollination? 18

1.7.1 Cross-pollination 19

1.7.2 Availability of pollinating insects,, ... 19

1.8 Palm Fruit Crops 20

1.8.1 Oil Palm 21

1.8.2 Coconuts 22

1.9 Local Tropical Fruits 23

1.9.1 Pineapple 23

1.9.2 Rambutan and other fruit crops 24

1.10 Corn / Maize, Sorghum & other cereals 25

1.11 Summary: 27

Chapter 2 28

2.0 Pollination efforts in the region – Example crops favourable to Stingless Bees 28

2.1 Cash Crop compatibility with Meliponiculture 28

2.1.1 Objectives 28

2.2 Issues about Providing a Pollination service as part of Meliponiculture development 28

2.3 Universal Protocols & Standard Operating Procedures (SOP) 29

2.4 Meliponines & Flower distinctiveness with Crop Farmer acceptance and agreeable rates 29

The individuality of Meliponines & Flower 29

Distinctiveness 29

..... 29

2.5 Crop Farmer Acceptance 30

2.6 Acceptable fees 30

Agreeable rates 30

2.7 Govt & Research Agency support – Pollination Seminar 31

2.8 Pollination Initiative 31

2.8.1 Predicaments faced by Beekeepers... 31
2.8.2 Predicaments faced by Crop Farmers 32
Chapter 3 34
3.0 Crop pollination data on Stingless Bees 34
List of crops favoured by Stingless bees 34
3.1 Avocado (*Persea americana*) 35
3.2 Banana (*Musa* spp.) 36
3.3 Brinjals - Eggplant (*Solanum melongena*) 37
3.4 Cocoa (*Theobroma cacao*) 38
3.5 Cashew Nut (*Anacardium occidentale*) 39
3.6 Chestnut (*Castanea sativa, C. mollissima*) 40
3.7 Chillies - (*Capsicum* spp.) 41
3.8 Citrus (*Citrus* spp.) 42
3.9 Coconut (*Cocos nucifera*) 43
3.10 Coffee (*Coffea* spp.) 44
3.11 Cucurbits: Cucurbita (*Cucumis* spp.) 45
3.12 Dragon fruit - Pitaya (*Hylocereus* spp.) 46
3.13 Durian – Durian Lai (*Durio zebethinus, D. kutejensis*) 47
3.14 Grapes (*Vitis vinifera*) 48
3.15 Guava (*Psidium guajava*) 49
3.16 Langsat (*Lansium domesticum*) 50
3.17 Longan (*Dimocarpus longan*) 51
3.18 Lychee (*Litchi chinensis*) 52
3.19 Jackfruit (*Artocarpus heterophyllus*) & Breadfruit (*Artocarpus altilis*) 53
3.20 Maize (*Zea mays*) 54
3.21 Mangosteen (*Garcinia mangostana*) 55
3.22 Mango, Kuini, Bacang (*Mangifera* spp.) 56
3.23 Oil palm (*Elaeis guineensis*) 57
3.24 Papaya - Pawpaw (*Carica papaya*) 58
3.25 Passion Fruit - Markisa (*Passiflora edulis*) 59
3.26 Pineapple (*Ananas comosus*) 60
3.27 Pomegranate (*Punica granatum*) 61
3.28 Pomelo & Grapefruit (*Citrus* spp.) 62
3.29 Rambutan (*Nephelium lappaceum*) 63
3.30 Roselle (*Hibiscus sabdariffa*) 64
3.31 Soursop & Annona (*Annona muricata, A. squamosa*) 65

3.32 Starfruit - Carambola (*Averrhoa carambola*) 66
3.33 Rose Apple, Jambu (*Syzygium jambos*) 67
3.34 Strawberry (*Fragaria* spp.) 68
3.35 Sunflower (*Helianthus annuus*) 69
3.36 Snakeskin Fruit - Salak (*Salacca* spp.) 70
3.37 Sapodilla – Ciku/Sawo (*Manikara zapota*) 71
3.38 Tamarind (*Tamarindus indica*) 72
3.39 Tomatoes (*Solanum lycopersicum*) 73
3.40 Watermelons (*Citrullus lanatus*) 75
Chapter 4 .... **79**
4.0 Meliponiculture of *Geniotrigona thoracica* (Smith 1857) in Malaysia 79
4.1 Abstract 79
4.2 Introduction 79
4.3 Materials and Methods. 80
4.3.1 Material Study and Placement in the Meliponary.... 80
4.4 Beescape in *G. thoracica* Meliponary 81
.... 82
4.5 Nest Structure 82
*4.5.1 N*est entrance tubes 82
4.5.2 Nest Structure in a feral hive (Figure 15) 83
4.5.3 Nest Structure within a boxed hive (Figure 16) 84
4.6.0 Failures due to weak colonies 85
4.6.1 Examples of taken over nests 86
4.7 Morphology - Comparisons between *Geniotrigona* and *Wallacetrigona* 87
4.8 Honey production support trials 88
4.8.0 Results and discussion 89
4.8.1 Meliponiculture 89
4.8.2 Intervention trials with artificial storage pots 89
4.9 Conclusion 91
Chapter 5 .... **93**
5.0 Concoctions of Meliponicultured Honey and Traditional Herbal Medicines Developed from Correlations with Ethnoentomology in Langkawi, Malaysia 93
5.1 Meliponiculture in Langkawi, Malaysia 94
5.2 Taman Kelulut Kompleks Kraf Langkawi 101
5.3 Some traditional practices that use honey 103
5.4 Traditional & Indigenous knowledge Vs Scientific Analysis & Findings 104
5.5 The MGI team for Data Analysis of the Honey concoctions from Langkawi Malaysia 104

5.6 Research Aim 104
5.7 Activities & sampling 106
Chapter 6 **107**
6.0 Unique Stingless bees of Penang Island, Malaysia 107
6.2 *Tetragonilla atripes* (Smith, 1857) 113
6.3 *Tetragonula fuscobalteata* (Cameron, 1908) 114
6.5 *Tetragonula pagdeniformis* (Sakagami, 1978) 117
6.6 *Tetragonula zucchii* (Sakagami, 1978) 118
6.7 *Tetrigona apicalis* (Smith, 1857) 119
6.8 *Tetragonula iridipennis* (Smith, 1854) 120
6.9 *Tetragonula melina* (Gribodo, 1893) 121
6.10 *Tetragonilla collina* (Smith, 1857) 123
6.11 *Lepidotrigona terminata (*Smith, F., 1878*)* 123
6.12 *Heterotrigona bakeri* (Cockerell, 1919) 124
Chapter 7 **127**
7.0 Historical records of Highland stingless bees in Malaysia 127
7.0.1 Records of *itama* from Schwarz 1939 127
7.0.2 Records of *erythrogastra* from Schwarz 1939 128
7.0.3 Records of *canifrons* from Schwarz 1939 128
7.0.4 Records of *fimbriata* from Schwarz 1939 129
7.0.5 Records of *thoracica* from Schwarz 1939 129
7.0.6 Records of *iridipennis* from Schwarz 1939 130
7.0.7 Records of *valdezi* from Schwarz 1939 131
7.0.8 Records of *geissleri* from Schwarz 1939.. 131
7.0.9 Records of *atripes* from Schwarz 1939 132
7.0.10 Records of *collina* from Schwarz 1939 132
7.0.11 Records of *fuscibasis* from Schwarz 1939 133
7.0.12 Records of *rufibasalis* from Schwarz 1939 133
7.0.13 Records of *apicalis* from Schwarz 1939 133
7.0.14 Records of *binghami* from Schwarz 1939 134
7.15 Records of *melanoleuca* from Schwarz 1939 134
7.0.16 Records of *peninsularis* from Schwarz 1939 135
7.0.17 Records of *nitidiventris* from Schwarz 1939 135
7.0.18 Records of *trochanterica* from Schwarz 1939 135
7.0.19 Records of *flavibasis* from Schwarz 1939 135
7.0.20 Records of *terminata* from Schwarz 1939 136

7.0.21 Records of *latebalteata* from Schwarz 1939 ... 136
7.1 Cameron Highlands - a) *Tetragonula geissleri* (Cockerell 1918) ... 137
Cameron Highlands - b) **Melipona basimaculata* Bingham 1903, nomen dubium. ... 137
7.2 Larut Hills - *Tetrigona apicalis* (Smith, 1857) ... 139
7.3 Mt Dulit - ... 140
7.3.1-Lowland Slopes ... 140
7.3.2 Mt. Dulit, At Elevation 2500 Feet ... 141
7.3.3 Mt. Dulit, At Elevation 4000 Feet ... 141
7.3.4 R. Kapah, Tributary of R. Tinjar ... 141
7.3.5 Mt. Kalulong, At Elevation 1800 Feet ... 141
7.4 Batang Padang, Tg Malim - *Tetrigona peninsularis* ... 142
7.5 Sungai Pinang Hill - *Tetragonula iridipennis* ... 142
7.7 Mt Ophir (Gn. Ledang) ... 142
7.7.1 Expedition ... 143
7.8 Fraser's Hill ... 148
7.9 Kancing Hill, Selangor ... 148
7.10 Bukit Ketam, Tuaran, Sabah ... 148
7.11 Kedah Peak (Gn Jerai) ... 149
7.12 Bukit Kutu, Selangor ... 149
7.13 Ulu Gombak- Forest Reserve ... 149
7.14 References ... 150
List of NHUK Museum specimen labels. ... 150
List of Illustrations ... 150
Chapter 8 ... 151
8.0.1 Introduction ... 151
8.0.2 Photo Gallery ... 151
8.1 Diversity of Stingless bee Species in Ranau, Sabah ... 154
8.1.1 Genus *Geniotrigona* ... 154
8.1.2 Genus *Heterotrigona* ... 155
8.1.3 Genus *Lepidotrigona* ... 156
8.1.4 Genus *Tetrigona* ... 157
8.1.5 Genus *Tetragonilla* ... 157
8.1.6 Genus *Lophotrigona* ... 158
8.1.7 Genus *Sundatrigona* ... 159
8.1.9 Genus *Tetragonula* ... 160
8.1.10 *Odontotrigona haematoptera* ... 162

8.1.11 Isolated Specimens 162
8.1.12 Unidentified Types 164
List of Fig in chapter 8 165
Chapter 9 166
9.0 Merlimau Kelulut Honey Farm 166
9.1 Introduction 166
9.2 Establishment 166
9.3 Production 168
9.4 Current Status 169
9.5 Aspirations 170
Chapter 10 172
10.0 Development of Meliponine Repositories in The Fraser Hill UKM Research Centre And 'Kelulut' Species Identification through DNA Barcoding and Comparison of Honeys 172
10.1 Introduction 172
10.2 Materials and Procedures 173
10.2.1 Development of the Meliponine Bukit Fraser Investigation Centre Repository 173
10.2.2 Sample Collection 174
10.2.3 Species Identification of Meliponines and Specimen Preparation 174
10.2.4 Extraction and Linkage of Meliponines DNA 174
10.2.5 DNA Reference Data Processing and Meliponine Phylogeny Analysis 175
10.3 Results and Discussion 176
10.3.1 PPBF Meliponine Repository Development 176
10.3.2 Sampling of meliponines for barcode analysis 176
10.3.3 Principal phylogenetic analysis 178
10.3.4 Comparison of Meliponine Honey Metabolite Profiles before and after Adaptation to PPBF 183
10.4 Conclusion 184
10.5 References 185
Chapter 11 187
11.0 Meliponine Repository at Malaysia Genome Institute (MGI) 187
11.0.1 Introduction.. 187
11.1 MGI Core Facilities 188
.... 189
11.6.1 Previous packing 194
11.6.2 New packing 194
11.7 SDS PAGE GEL11.8 Application of DNA Barcoding in Identifying Stingless Bee Colonies - MSI 16046 (2016) 195
11.9 DNA Barcoding 198

Chapter 12 199

12.0 Meliponiculture & Nature Therapy for Learning Disorders Kid@Ukm Fraser Hill Research Centre 199

12.1 Abstract 199

12.2 Introduction 200

12.3 Materials and Methods 201

12.3.1 The Program Itinerary 201

12.3.2 Event highlights in the program 202

12.3.3 Facilitators 202

12.3.4 Participants 203

12.4 Results and Discussion 203

12.4.1 Feedback Facilitators 203

12.4.2 Feedback Participants 204

12.4.3 Questionnaire (Post Event) for Feedback and Evaluation 204

12.5 Compilation of pictures 207

12.6 Conclusion 208

12.7 Acknowledgements 209

12.8 References 209

Chapter 13 210

13.0 Diversity in Stingless Bee Honey Profiles and Implications of Trehalulose Discovery as One of The Main Sugars in Stingless Bee Honey 210

13.1 Abstract 210

13.2 Introduction 210

13.3 Difference between stingless bee species that leads to the production of different honey quantities and qualities 211

13.3.1 Species of stingless bees 211

13.5 Physicochemical properties of some stingless 212

13.5.1 Comparison of stingless bee honey composition based on the species, origin and nectar preferences 212

13.5.2 Discovery of trehalulose sugar 213

13.5.3 Inaccurate reports of maltose in stingless bee honey 214

13.6 Tabulations 215

13.7 Conclusion 227

Chapter 14 232

14.0 Stingless Beekeeping Practices and Honey Industry From A Malaysian Perspective 232

14.1 Introduction 232

14.2 History and Beekeeping Practices in Malaysia 233

14.3 Honeybees and Stingless Bees 234

14.4 Bee Products 234
14.5 Honey Industry in Malaysia 236
14.6 Sustainable Beekeeping and Challenges 237
14.7 Future Outlook 238
14.8 Reference 239
Chapter 15 240
15.0 Learning from Failures in Meliponiculture and Improper Strategies of Stingless Beekeeping in Malaysia 240
15.1 Synopsis 240
15.2 Introduction 241
15.3 The Meliponiculture Industry Is a New Industry 241
15.3.1 Lack of information 241
15.3.2 Procurement – The type or nature of the colonies taken and removed from the habitat of origin: 242
15.3.3 Felling / collecting colonies at the wrong time 243
15.3.4 Getting colonies from across borders 244
15.3.5 Smuggling logs from neighbouring countries. 245
15.3.6 Incorrect cutting procedure 245
15.3.7 Acquired colony at daytime - Transfer of colony during the day/heat 245
15.3.8 Logs in Transit 246
15.4 Problems faced when farming in urban or suburban areas 250
15.4.1 The use of herbicides and mosquito fogging 251
15.4.2 Bee deaths due to herbicides/insecticides. 251
15.4.2.1 Methods of overcoming mosquito fogging 251
15.4.3 Improper strategies in breeding farms 252
15.4.4 The proposed volume of hive boxes 253
15.4.5 Diverse selection 253
15.4.6 Exposure of paints in beekeeping 254
15.5 Species Distribution 256
15.6 Dipterocarpaceae in Malaysia - Endemic Resin producing Trees 257
15.6.1 Common resinous trees with botanical and vernacular names. 259
Chapter 16 267
16.0 The Itama Journey 267
16.1 The Journey Log 268
16.1 List of Locations 269
Results and Discussion 297
Chapter 17 298

17.0 Advances in Beescape for Meliponines 298
17.1. Plantings in 4 strata levels 298
17.1.1 Resource Partitioning 298
17.2. Communication with the bees in Beescape planning 299
.... 299
17.3. Meliponary Topsoil, Relative humidity and Groundcovers 300
17.4 Decaying n decomposing organic matter 301
17.6 Windbreakers n perimeter planting 302
17.7. Managing swarms in a Meliponary 302
17.8. Foraging peculiarities 303
17.9. Landscape jargon in the ecological context 305
Bibliography 307
List of Figures 312
Appendix A - Glossary on Pollination terminology 315
Appendix B – Genera Identification Key Update 333
Appendix C – Types of swarms 334
Index 337

Foreword I

Congratulations to Mr. Abu Hassan Jalil (Pak Abu) as the coordinator and Dr. David W. Roubik as Editor for the publication of the second series of books entitled "Meliponiculture and Beyond", specifically on the development of Meliponiculture in Malaysia. It is an impressive effort to document 'kelulut' beekeeping activities for the eyes of those involved in this field and those who are interested in following its development.

Meliponiculture and 'kelulut' honey are no strangers to the community in Malaysia. Since the introduction of 'kelulut' farming in 2010, it has become the talk of the town in urban and rural areas, among farmers, entrepreneurs, consumers and even national leaders. Efforts to develop meliponiculture are now having a major impact on pollination as well as honey production. Efforts are made in flatland agricultural areas and forest edges, and upland areas such as Cameron Highlands.

Farmers are focused on increasing quantity and are now encouraged to practice quality breeding (My GAP) and honey production that meets standards (Honey Standard).

This book looks at what has been done concerning the successes and failures of the past. It also invites us to look ahead to what can be developed to enhance further the ability of entrepreneurs and the variety of uses of kelulut from pollination to products such as honey, pot-pollen and propolis.

Congratulations to all the writers who have successfully documented the beekeeping work, 'kelulut' research and Malaysia's results. Overall, this is a very proud moment and can be a stimulus for further progress in livestock and agricultural development and the production of 'kelulut' products in the future.

SITI AMRAH SULAIMAN (PROF. PhD)

Foreword II

Beekeeping products are potential sources of additional income for poverty eradication in tropical countries. Having researched beekeeping for more than 30 years, I have read many books on Apiculture and Meliponiculture. The latest book by Mr. Abu Hassan Jalil, a Meliponiculture industry expert, should contribute immense knowledge to the industry development in Malaysia and other tropical countries. The book is beneficial to players in the beekeeping industry supply chain, researchers and policymakers worldwide. Even though Meliponiculture was recently introduced in Malaysia, its potential as a new export commodity to enhance economic growth is inevitable. This book sets an array of policies that would pave the way to the future development of the Meliponiculture industry in tropical countries. Pollination is the main product of beekeeping.

Perfect pollination is a source of sustainable, cleaner production where food crops productivity increase without the additional use of hazardous inputs like pesticides and fertilizers. Honey, another product of Meliponiculture, can increase its production by way of integrating current beekeeping with highland bee farming. Also, increase the population growth of stingless bees. As described in the book, specific species in both areas contribute towards pollination and honey production differently.

The book explores the honey quality and sugar content and its potential as diabetic food to increase the honey demand. Better honey quality from the proper standard operating procedures will boost the market for stingless bee honey and fetch premium prices. In summary, the book is beneficial to individuals venturing into Meliponiculture and the Government interested in developing the industry to eradicate poverty among rural and urban poor.

MOHD MANSOR ISMAIL (PROF. PhD)

Preface

In the grand scheme of things, this book explores the central theme of stingless bees' importance and utilisation.

Sustainable development in everyday agriculture instead of advanced technically innovative like aeroponics or molecular growth enhancers, flowers need to be pollinated. Regardless of pollination variations, flowers and bees are part of nature and agricultural produce is part and parcel of standard fertilization and crop reproduction.

Utilizing specific pollinating agents like native insects for native crops would be the most appropriate effort in crop yield management.

With expanding urban development and environmental pollution on the rise, human civilization and industrialization take over agricultural space. It seems prudent for humans to revert to higher altitude agrarian development.

Meliponiculture being a growing trend after the current decline of apiculture and monoculture farming is pushed towards the suburbs and eventually to higher ground. This book looks at the potential of highland meliponiculture in the ever-increasing highland agricultural development.

A view of historical records of high-altitude discovery of stingless bees shows a good environment for stingless bees conservation and utilization in highland agriculture. As an infant step in conservation efforts, there is a need to address past failures in Meliponiculture and upgrade pollination goals with stingless bees rearing in the highlands.

The subsidiary produces in stingless bee utilization in agriculture justifies the need for more major efforts to conserve our national treasure. A chapter on the analysis of diverse sugar profiles gives great enlightenment on the importance of honey derived from meliponines.

The book ends with the journey of honey sampling at Meliponaries throughout Peninsular Malaysia.

Chapter 1

1.0 Pollination Initiatives in Development of Meliponiculture with Cash Crop Agriculture

By Abu Hassan Jalil

Akademi Kelulut Malaysia Sdn Bhd.

1.1 Abstract:

Stingless bees in pollination of economic and cash crops are highly valued. Being native to Asia, stingless bees are more resistant to pests and disease, sustainable and more efficient pollinators for native plant species compared to introduced foreign bees. Several technologies are developed to manage stingless bees for crop pollination and honey, pollen, and propolis production. However, the compatibility of meliponiculture with certain cash crops is questionable. Seminars were conducted to generate a collaborative model for pollination of crops utilizing stingless bees among crop farmers and stingless beekeepers. Other initiatives were to formulate a possible solution towards a profitable business model for a win-win situation for both crop farmers and stingless beekeepers. Besides this, the objective was to introduce and encourage similar activities in both parties, i.e. Beekeepers as crop farmers and crop farmers to incorporate stingless beekeeping.

Meliponiculture with Pineapple farming is a fallacy because Pineapples (*Ananas comosum*) have their flowers of the unpollinated plant fuse to form "multiple fruits" (multiple flowers each with one ovary that produce one berry, which later merges with its neighbouring berries to form one pineapple) (Morton J., 1987). Anyway, the monoculture practice on large acreage of Pineapple farms is not suitable for meliponiculture.

Meliponiculture is more suited and ideal for palm crops like coconuts and betel nut. Other palm crops that are bee pollination dependent are Arenga nut and Snakeskin fruits. Other cash crops like self-fruiting bananas, wind-pollinated nut-producing trees or nocturnal flowering crops like the durian (*Durio zebethinus*) and the dragon fruit (Pitaya) may not need the presence of stingless bees for yield improvements (Mabberley 2017). This paper looks at cash crop farming that is suitable for Meliponiculture activities.

Keywords: Cash Crops, Meliponiculture compatibility, Pistillate Anthesis, Staminate Anthesis Nocturnal Flowering, Insect Pollination.

1.2 Introduction

Plant pollination is a clear example of ecosystem services - economic benefits afforded to nature (Millennium Ecosystem Assessment 2005). Bees and other pollinating agents provide valuable inputs to agricultural production that can increase the harvest's size and quality (Allen-Wardell et al., 1998). Plants such as tomatoes, coffee, canola, melon, and cocoa increase animal pollination (NRC 2007). Meliponines benefited from plant breeding efforts using pollen, nectar, gum or latex and oils (essential oils as well) from the flowers, and using phloem, resin and extrafloral nectar of floral nectar and sap, which formed the plant defences against herbivores (Roubik 1993).

In 2008, MARDI (Malaysian Agriculture Research & Development Institute) introduced a cash crop pollination business using stingless bees. (Conservation and sustainable utilization of stingless bees for pollination services in agricultural ecosystems in Malaysia, Norowi et al., 2008). At the time, the value of pollination services using insects for star fruit, guava, lime, mango, watermelon, durian, and coconut, is estimated USD 19 million a year (Norowi et al., 2008).

Among the insects identified are four species of stingless bees, i.e. *Heterotrigona itama, Geniotrigona thoracica, Tetragonilla atripes* and *Tetrigona peninsularis*. In 2010, Dr. Ibrahim Shrub (MBBS) and the authors tried to apply this pollination service through their "Great Farm" farm in Kg. Sg. Jang, Kerling, Hulu Selangor. The fruits found in the fields are star fruit, guava, lemon, mango, durian, and coconuts. Various species of meliponines, including five species recommended by MARDI, *Heterotrigona itama, Geniotrigona thoracica, Lepidotrigona terminata, Tetragonula laeviceps* and *Tetrigona apicalis.*

1.3 Initial Trials & Obstacles

Experiments and research on a pollination service business cultivated among our cash crops in the field were carried out. The first hindrance encountered was proper farming for the bee species diversity (there were 11 species in the area at the start of the venture).

Experiments carried out on the design and size of portable box hive dimensions for pollination efforts. Some setbacks encountered were as follows:

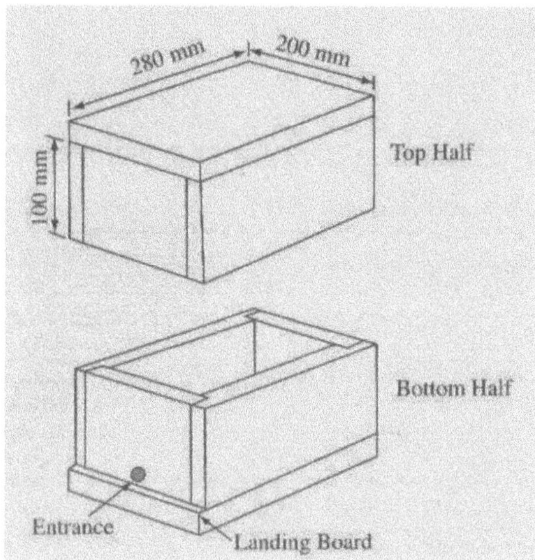

Figure 1 Typical portable box hive for pollination

1. The obstacle is to provide a universal "template" for the known species because different species will differ in size. The pollination services requirements vary as every species colony will have a diverse average population.
2. Various types of hardwood were also crucial for the colony's development and sustainability to produce durable boxes.
3. The flower physiology and duration of fruit flowers in bloom vary drastically. Stingless bees have variable attraction affinity for a variety of flower shapes and colours.
4. In addition to the above, attempts were made to conduct pollination programs with other crop farmers, such as nearby rock melon entrepreneurs. Unfortunately, many of them are unaware of or less likely to try this pollination service because of the worries of damage to their products. They also do not readily accept any changes in the pesticide usage schedule in their maintenance.
5. Proposed rental rates (RM 20.00 a box hive hired for a duration of 7 to 10 days depending on the crop) are considered expensive as there have been no proven data (effectiveness in the field) at that time.
6. The breeding service business could not respond because the information suggested by MARDI was not widely distributed to farmers. In 2012, MARDI had changed its direction, rather than focusing on only pollination, and it decided to concentrate on Melponine Honey's production.

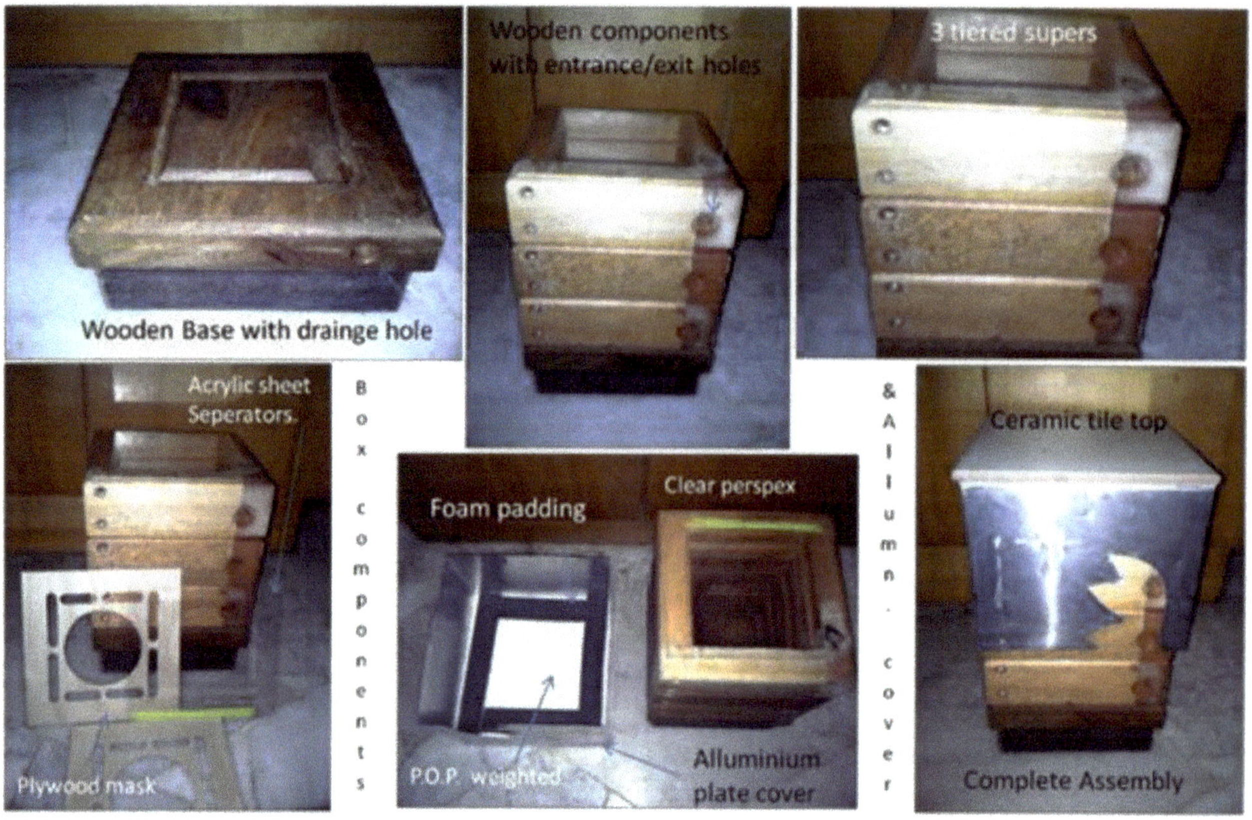

Figure 2 Durable 3-tiered box hive

Box design and meliponiculture systems also change to a beekeeping system that uses the "Topping" box rather than the portable pollination box. With this, the pollination services in Malaysia deteriorated.

In the Philippines, pollination services are still being accepted in Mango farm fields, Coconut Farm fields and Pili farm fields with a density of one nest for four mature plant mature trees. There is also a lease system of 3,000 pesos per *Tetragonula biroi* box colony (C. Cervancia & A. Fajardo, personal communication, Oct. 2012)

1.4 Recommendations

In 2011, the total value of RM 1.3B fruit consumption was calculated by the average pollinator insect's dependence @ 0.65 = RM845M. Based on Plant Pollinator Interaction charts in Momose et al.1998, social bees make up 36% of the total number of insect pollinators. This means that IPEV (Value of pollination insect dependence) for social bees is RM845M x 36% = RM 303M in 2011.

Encouraged by the above data, the pollination service business resumed by holding seminars or information dissemination sessions from time to time. These seminars prove to be fruitful when the representatives of the cash crop farmers are gathered together with meliponiculture farmers for discussions and exchange of recommendations.

1.5 Pollination Initiative Seminar

Objectives:

To initiate a collaborative model for pollination of crops utilizing Stingless bees between crop farmers and beekeepers

A) To formulate a possible solution towards a profitable business model for a win-win situation for both crop farmers and beekeepers.

B) To introduce and encourage similar activities in both parties, i.e. Beekeepers as crop farmers and crop farmers to incorporate stingless beekeeping.

C) To resolve past issues and overcome retarding factors & effects in stingless bee utilization in crop pollination.

Issues as follows:

1. There is the extensive usage of pesticides by crop farmers. This practice means Crop farmers have to reschedule their maintenance regime.
2. The business model was not agreed upon amicably (e.g. who pays whom how much concerning increased yield)
3. Beekeepers do not have proper mobile pollination boxes. Fabrication of new packages with new templates for different species will incur further expenses.
4. Distance and transportation from beekeeper to crop farm became a negative factor. Time and duration of colony movement became an issue with new labour and pedestals for placement.
5. The long-term effects for bees in a "greenhouse" environment are re-planning and redesigning "greenhouse" features needed to incorporate bees for a longer duration.
6. Frequent to and fro movement of hives from crop farm to bee farm may cause colony development issues (this may increase meliponiculture costs)
7. Crop farmers are not willing to pay for hive rental without field evidence—another form of compensation quantification is required and must be agreed upon by all parties.
8. Which party bears the cost of colony failures? Was the hive weak at entry in the first place, or was there some parameter in the farm that caused colony failure? Colony health at

the start must be determined and agreed upon. Parameters at crop farms must be suitable for colony entry.

9. Who determines and monitors the rate and type of pesticide usage allowable? MyGap organic farming or EM usage be defined and adhered to.

10. In the early 2000s, a farm experienced an overwhelming increase in yield brought about by *Apis* pollination. The management increased their order of fruit packaging, and that cost increase was charged to the participating beekeeper! Crop farmers must bear all incurring costs that come with growing yields and hence increased profits. How can the beekeeper also benefit from this profit increase (if any)?

1.6 Indications of crop benefits from insect pollination

We explore indications that the following tropical crops benefit from insect pollination: allspice, avocado, cashew, coconut, coffee, cotton, lychee, mango, melon, passion fruit, safflower, sunflower, and many varieties of *Citrus* and *Cucurbita*.

Likely, some safflower, mango, okra, and opium poppy (*Papaver somniferum*) give only moderately increased yield when pollinated by insects while others (for example, cashew and guava) offer significantly increased profits. However, more research is needed. Other crops' pollination requirements, including tropical legumes, which are beginning in large-scale production, are mostly unknown.

Because of the lack of time and facilities, too many studies on tropical crop pollination are preliminary. Careful investigations instigated to determine:

a) whether fruit or seed set is adequately improved;

b) whether the crops benefit from pollination; and

c) which insects pollinate the flowers.

1.7 Does a crop need insect pollination?

A standard method to determine whether a crop benefit from insect pollination is to compare the yields of plants that are grown under three different conditions:

a) covered by nylon screen cages containing bee colonies;

b) protected by barriers to exclude insects;

c) not caged.

If cages are not available useful information is obtained by bagging individual flowers or flower heads to exclude pollinating insects. Some of the bagged flowers must be hand-pollinated to find the maximum set possible under these conditions.
It was determined whether the crop yields are resultants of self- or cross-pollination.

1.7.1 Cross-pollination

Cross-pollination is needed when the sexes are segregated on different plants (for example, melon) or by different periods of the same plant's flowering (for example, avocado). Cross-pollination may also increase the yields of plants that can be self-pollinated. Cross-pollination can only occur when the pollinating agent moves from one plant to another, and this usually happens on only a small proportion of flower visits.
Production of hybrid seed crops on a commercial scale creates a particular need for cross-pollination by insects. A high population of pollinating insects is needed to carry pollen from rows of male plants to rows of female plants. Many varieties of fruit trees need cross-pollination. When an insect moves from a pollinizer tree to a tree of the main array, it pollinates only the first few flowers it visits. As such, parts of trees adjacent to that pollenizer tree get the most fruit set. For an even fruit set, orchards plan such that pollen providing or pollenizer trees surround the main variety of trees.

1.7.2 Availability of pollinating insects

Having considered the need for pollination, I wish now to turn to the availability of pollinating insects. Probably humans at first encouraged wild bees and other pollinating insects as primaeval forests cleared, allowing light to penetrate and low growing plant species to multiply. However, in many parts of the world, including the tropics, there is a tendency for populations of wild bees and other pollinating insects to diminish until they are too few for crop pollination.

Several factors have contributed towards this:

1. Intensive cultivation of the land, including the lessening of hedgerows, banks and rough verges, has diminished the number of nesting and aestivation or resting sites;

2. Regular cutting of roadside verges and around field crops and the use of herbicides has destroyed flowering plants that help to provide beneficial insects with food;
3. The use of selective herbicides destroy flowering weeds within crops;
4. Use of insecticides. These are probably the most severe threat to wild pollinators in agricultural areas and provide the most significant deterrent to using honeybees as pollinators.

The damage to honeybee colonies differs according to many factors, including the insecticide's toxicity, the method and time of day of application, the number of applications, the proportion of foragers visiting the treated crop, the crop and its floral structure, the bees foraging on the harvest, and the 'drift' of insecticide to other sources of forage. Knowledge on the toxic effects of insecticides on wild bees and honeybees is lacking in the tropics, although some information was now gradually available.

Fortunately, practices tending to diminish wild bee populations are in general less severe at present in the tropics. But unless care is taken to avoid destroying wild pollinators in tropical countries, pollination will be almost entirely dependent on *Apis* bees. Already, it is supposed that in several tropical countries, the honeybee bee colonies managed by beekeepers will soon be the primary pollinators. However, in the past, large areas of vacant bushland and scattered holdings provided ample nest sites for wild pollinators, including wild honeybees.

Modern agricultural practice tends to diminish the number of wild pollinators and increase the demand for pollinating insects.

1.7.4 Cash Crop Compatibility

When an administrative mindset looks at major crops, their first thought is pollination and meliponines. Integration program of a rubber plantation for meliponines because a) you get sunlit pockets for flowering shrubs during rubber flower offseason and b) the rubber tree provides sufficient meliponines nest-building material. This compatibility does not apply to many other plantation crops. The technical mindset must assist the administrative!

1.8 Palm Fruit Crops

Table 1 List of palm fruit crops with vernacular names

	Botanical Name	Palm Nut Crop	Vernacular name
1	*Areca catechu*	Betel Nut	Buah Pinang
2	*Arenga pinnata*	Arengga nut	Buah Aren, Buah Enau, Kolang Kaleng

3	*Bactris gasipaes*	Peach Palm	Chontadura (Spanish)
4	*Borassus flabellifer*	Palmyra fruit	Kelapa laut / Toddy Fruit / Nungu
5	*Butia capitata*	Jelly Palm	Pindo (Spanish)
6	*Cocos nucifera* var.	Coconut varieties	Kelapa pandan, K. gading & K. Matag
7	*Elaeis guineensis*	Oil Palm	Kelapa sawit
8	*Euterpe oleracea*	Acai berry / Asai Palm Fruit	Buah Acai / Asai
9	*Phoenix dactylifera*	Date Palm	Buah kurma
10	*Salacca zalacca*	Snakeskin fruit	Buah Salak

1.8.1 Oil Palm

As for oil palm, the plant's 30' canopy covers everything and does not support ground cover or shrubs for the bees' resource partitioning; there is practically no propolis source. However, in nature, oil palm is not pollinated by bees.

The pollen of oil palm (*Elaeis guineensis* Jacq.) is a potent allergen in Malaysia and Singapore (Kimura, 2003). 14.8% of the total of 32.86 million hectares of Malaysia's total land area (Thang, 2005), Ministry of Agriculture, Department of Statistics, Malaysia. Of this, 50% of Stingless Beekeepers are estimated to practice Meliponiculture within or at the fringe of Oil Palm Estates throughout Malaysia.

Therefore, this brings rise to a common question "How would this affect asthmatic prone individuals and hay-fever sufferers consuming honey or "pot-pollen" produced from such areas (Oil Palm plantations)?"

Besides the oil, other commonly edible Tropical Palm fruits are as in the above (*Table 1 List of palm fruit crops with vernacular names*). These palms have elaborate inflorescence in each spadix and are popular choices for stingless bees. The stingless bees are good pollinators for palms and produce excellent yields. The coconuts are described separately below. All the palms thrive in tropical weather, and *Butia capitata,* even though native to South America, is notably exceptional as one of the hardiest feather palms, tolerating temperatures down to about −10 °C; widely cultivated in temperate climates. Pollinators were bees, wasps, flies, and beetles. (Henderson, 1986)

Betel nuts are quite popular in the Indian sub-continent and Sri Lanka, while the Arenga nut and *Salacca* spp and varieties are widely cultivated in the Indonesian archipelago. The

Palmyra fruits are grown for Toddy production in S.E. Asia and India, while Date Palms, Peach Palms and Asai Palms are currently cultivated in South East Asia.

1.8.2 Coconuts

Cocos nucifera palms appear to be most compatible with Meliponiculture. Coconut pollen grains are monocolpate. Those of the variety typically measured 0.065 to 0.069 mm. in length and 0.028 to 0.030 mm. in diameter. The pollen output per anther in a healthy flower is estimated to be between 111,000 and 221,000 grains. (Manthriratna, MAPP, 1965)

Image 1 Among the foraging favourites across all the species of Meliponini

Cocos nucifera is monoecious, protandrous, and has unisexual flowers. Purseglove (1972) has described inflorescence development. A prophyll and peduncular bract enclose the inflorescence. These split abaxially, and the rachillae are exposed. Rachillae bear a few pistillate flowers at the base and hundreds of staminate (male) flowers above. Staminate flowers open in a distal to proximal direction. Anthesis of individual staminate flowers lasts one day, but staminate anthesis continues for 16-22 days. The pistillode produces nectar. Pistillate (female) anthesis follows and lasts from five to seven days. Anthesis also proceeds in a distal to proximal direction. Pistillate flowers have nectar. Usually, pistillate and staminate anthesis do not overlap.

Sholdt and Mitchell (1967) studied coconut pollination in Hawaii. They found staminate anthesis to last an average of 24 days and pistillate for an average of four days. Individual pistillate flowers were receptive for an average of three days. There is some overlap between phases observed and some overlap between successive inflorescences, both of which could cause self-pollination. Of 51 species of insect visitors to inflorescences, seven were common on both staminate and pistillate flowers. Sholdt and Mitchell concluded that honeybees were the primary pollinators, but also wind pollination occurred to a certain extent (Henderson, 1986).

Besides the nectar and pollen supply for Meliponines, fallen coconuts are also a haven for them. Old fallen fruits are usually pierced and foraged by squirrels leave behind a convenient nesting host for stingless bees (Jalil & Shuib, 2014). See below.

1.9

Image 2 Meliponine nests found in coconuts (previously bored by squirrels) in neglected coconut farms

Local Tropical Fruits

1.9.1 Pineapple

Pineapples (*Ananas comosus*) are self-infertile. They are parthenocarpic in that it a fruit still develops even with no pollination. Since hummingbirds are the principal pollinators, they are prohibited in Hawaii to avoid developing undesired seeds in their pineapples (Morton J., 1987). https://www.hort.purdue.edu/newcrop/morton/pineapple.html

Anyway, Meliponines and pineapple pollination is a fallacy! *Apis* bees (having a good flight range) gain from the fact that it takes nectar without providing any help to pineapples' pollination. Meliponines rearing at pineapple plantations is not a good practice because of the monoculture. Stingless bees will not have the opportunity to forage many needs because the flight range is limited, especially if placed in the middle of a pineapple plantation! Meliponines in the middle of a vast pineapple plantation may not obtain sufficient material for nest building.

Image 3 Pineapples as far as the eye can see

Malaysian Pineapple Industry Board (MPIB) spokesman said that "The national **pineapple production** for **2020** is projected to see a 60% increase of 2018". To achieve that goal, the Government had planned to increase the pineapple plantation acreage by 7,120-ha by 2020. Sarawak recorded the largest pineapple plantation areas after Johor, with 1,805-ha currently.

(https://www.thestar.com.my/news/nation/2020/06/12/govt-to-expand-pineapple-cultivation-with-booming-global-demand)

"The 2,792-ha plantation, located in Simpang Renggam about an hour's drive from Johor Baru, yields around 60,000 tonnes of pineapple annually." The meliponines will have nowhere to go for the many other resources they need with such a vast area. As a testament to the negative effects of monoculture on meliponiculture, this is why meliponiculture cannot be encouraged in a pineapple plantation. https://www.thestar.com.my/news/community/2006/06/24/johor-still-tops-in-pineapple/#zX0mhxxZLmpApV46.99

1.9.2 Rambutan and other fruit crops

Seasonal fruit crops are appropriate with meliponiculture only if there is some form of integrated farming. For example, citrus fruits farming combined with mango varieties (*Mangifera* spp.), sapodilla (*Manikara zapota*) and rambutan (*Nephelium lappaceum*) provide for continuous forage for the stingless bees. Otherwise, when the flowers have yielded fruit, the bees are at a loss for forage, and the beekeeper faces the inconvenience and the dangers of failures.

Citrus fruits like Lime, Calamansi Lime, Kaffir Lime, Tangarine, Calamondin, Pomelo, and tropical oranges. These are great candidates for integrated plantings. Also, with *Mangifera* spp. i.e. mango cultivars (*M. indica*) and varieties like kuini and machang *(M. foetida*). Other fruits like the *Syzygium jambos* or varieties (*Syzygium* spp.), (*Psidium guajava*) and guava varieties, cashew nut (*Anacardium occidentale*) and nutmeg (*Myristica fragrans*) are widely cultivated in S. E. Asia and the Indonesian archipelago. All these show excellent yields when farmed with stingless bees.

1.10 Corn / Maize, Sorghum & other cereals

Corn and other cereals, or even sorghum, have abundant pollen that may be ideal for meliponines. (Image 4)[1] Then again, corn has a larger pollen size (95 μm) than the average (http://www-saps.plantsci.cam.ac.uk/pollen/index.htm).

This crop is a species with substantial pollen grains of 0.070 by 0.090 mm and egg shape. *Zea mays* have the largest pollen grains among grasses. The pollen grain has a pale yellowish colour. Maize pollen is too big to be travelling a long distance through the air. It falls from the male tassels on the long female silks of the neighbouring plant.

Ref: (http://www.vcbio.science.ru.nl/en/virtuallessons/pollenmorphology/).

Some detailed studies can be done on its grain sediments in honey[2]. This predicament may be of concern in the monoculture of vast areas of corn farms. It may affect the absolute clarity of the resultant honey produced and retard marketability.

Editor's notes:

[1] NOT POLLINATING. They remove pollen, probably with no negative consequence to the plant (which produces 100s to 1000s of times more pollen than needed, and bees have little or no conceivable contribution to pollination.

[2] grains in honey are widely used and also widely completely misinterpreted. They show pollen arrives in bee nests or on bees and their honey. It shows 1) nothing about pollination, and 2) nothing about honey sources, WITHOUT OTHER DATA AND VALID ANALYSIS.

Stingless Bees & Cereals

Image 4 Stingless bee visits on Corn and Cereals

1.11 Summary:

Changes in the pollination community[3] will affect the vegetation composition and the entire food chain. It is essential for activities related to forests, logging, and recommendations for best practices (My GAP – Good Agricultural Practice Standards for Beekeeping) with other beehives and management.

Institutional Research Institutions and Institutions of Higher Learning continue to focus and research on pollination efforts and ensure that this research is widely disseminated, especially among the Association of Beekeepers, Cooperatives and Government-linked farmers' unions in the country.

Climate conditions are unpredictable in a fluctuating tropical environment, which may affect the colony's survival. This subject should be considered and monitored as a long-term activity and how different cultures (agriculture, horticulture & meliponiculture) adapt.

There is a need for multiplication of hives for sustainable pollination services with the following indicators to be taken into scrutiny:

1. Nest trap in some biomes is an indicator of which species will be for this method's adhesion.
2. To test several traps[4] nests and to observe which should be of enough use;
3. Standard methods using artificial baits to assess the presence of local species in the environment. Salt and sugar salts have been used[5] successfully (Roubik, 1996) in Asia.

Editor's notes

[3] Unless bees move pollen to an anther and it results in fruit and seed production, they are NOT pollinating.

[4] We do not want traps. We want artificial domiciles- or facilitated nesting. Traps usually are intended to snare and kill things- like bears, for example...

[5] applied as a spray on leaves to attract foraging bees.

Chapter 2

2.0 Pollination efforts in the region – Example crops favourable to Stingless Bees

2.1 Cash Crop compatibility with Meliponiculture

2.1.1 Objectives

- To observe the flower-visiting habits of stingless bees concerning cash crop production and yield influences.
- To determine the value of pollinating insect dependence.
- To ascertain the sustainability of meliponiculture in a yield improvement scenario.
- To introduce and encourage similar activities in both parties, i.e. beekeepers and growers who can benefit from incorporating stingless beekeeping reciprocally.
- To study the compatibility of cash crops agriculture with Meliponiculture activities.

2.2 Issues about Providing a Pollination service as part of Meliponiculture development

- **Challenges**: to provide a universal "template" for the widespread species.
- **Durable boxes:** hardwood types necessary for the development and sustainability of the colony. (added cost)
- **Pairing Flower and Bee**: flowering & duration of fruit flowers in bloom vary. Types of stingless bees also have different affinities for a particular kind[6] of flowers and colours.
- **Farmer acceptance**[7]: many are unaware and less likely to try this pollination service for fear of damage to their products.
- **Changes in Pesticide usage**: not readily accept any changes in the pesticide usage schedule in their maintenance.
- **Awareness**: The breeding services business could not respond because the information was not widely distributed to farmers by appropriate authorities.

---

[6] Editor's note: Some have studied a preference for a particular kind of flower or resource production in a flower patch. The rest is unknown.

[7] This is widespread and also explained and documented in the FAO book on pollination, two volumes, published in 2018. In Africa, for example, for the past 100 years or so the locals thought bee visits damaged flowers. That is 'traditional knowledge' in this part of the world. Imagine that! The real hesitancy we found in Mexico, which was justified, with growers, was that we were going to remove flowers and do harm with bagging or other manipulation, in doing our research project. Imagine that! However, since these were squash female flowers, each flower removed is one less squash produced for the farmer!

2.3 Universal Protocols & Standard Operating Procedures (SOP)

- **Objective**: The aim is to provide a universal "template" for the known species because different species will differ in size for additional pollination services requirements. Every species will have a diverse average population.
- **Varied affinity**: Different species have different preferences and different relationships for varied flowers and sap resources.
- Standards: To design and draw up appropriate operating procedures that can be as standards. (SOP)

2.4 Meliponines & Flower distinctiveness with Crop Farmer acceptance and agreeable rates

The individuality of Meliponines & Flower
Distinctiveness

- Duration: The flowering and period of fruit flowers in bloom vary drastically. Types of crops have different anthesis times and a variation of flowers and colours.
- Variations: Different plants and trees have varying understory requirements, such as humidity & temperature stability.
- Predators: Trees harbour and attract varied predators or enemies of Meliponines as well.

Figure 3 Meliponines & Flower distinctiveness

Figure 4 Cucurbit Pollination

2.5 Crop Farmer Acceptance

- Greenhouse operators: attempts to conduct pollination programs with greenhouse farmers were not ready to accommodate modifications.
- Awareness: Many of them do not try using this pollination service because of the worries of damage to their products.
- Pesticide Regime: not ready for any changes in their pesticide usage schedule in their maintenance.

2.6 Acceptable fees

Agreeable rates

- Rental rates: hive hiring for a limited duration is considered expensive by farmers.
- Records: no consistent pollination data records (effectiveness in the field) for many crops.
- Commitment: reluctant to invest upfront, and beekeepers do not want to transport their hives without initial monetary deposits.
- Binding: contracts seems too complicated for short term arrangements among both parties.

Figure 5 Honeydew melon trellis

2.7 Govt & Research Agency support – Pollination Seminar

- The pollination service business was not able to bloom because information suggested by Agricultural R&D Institutions was not distributed to farmers and
- In 2012, many agencies changed their direction rather than pollination, focusing on Meliponine Honey's production.
- The Agricultural Ministry did not pursue the Pollination services initiative because the 'organic' or pesticide-free policy is still in infancy.

Figure 6 Pesticide usage predicament

2.8 Pollination Initiative

Objectives:

- To initiate a collaborative model for pollination of crops between farmers and beekeepers to utilize Stingless bees fully
- Formulate a possible solution towards a profitable business model for mutual gains for both crop farmers and stingless beekeepers.
- To introduce and encourage similar activities in both parties, i.e. Beekeepers as crop farmers and crop farmers to incorporate stingless beekeeping.
- To resolve past issues and overcome retarding factors & effects in stingless bee utilization in crop pollination. Issues as follows:

2.8.1 Predicaments faced by Beekeepers

- **Unregulated pesticides usage:** Crop farmers have to reschedule their maintenance regime.
- **Business model** (e.g. who pays whom how much concerning increase in yield) How may the beekeeper also benefit from this profit increase (if any)?
- **Fabrication of new boxes** with new templates for different species will incur new expenses as beekeepers do not have proper portable hive boxes for the pollination service.

- **Distance and transportation** Time and duration of colony movement becomes an issue with new labour and pedestals for placement.
- **The sustainability of beekeeping** in a greenhouse environment ate detrimentally insufficient in nest-building materials forage source.
- **Added costs** due to frequent movement of hives may cause colony development issues (e.g. specially customized racks for towing the boxes).

2.8.2 Predicaments faced by Crop Farmers

Predicaments faced by Crop farmers

- **Revamping and redesign** of "greenhouse" features are needed to incorporate bees for a longer duration.
- **The reluctance is** of farmers to pay for beehive rental without field evidence.
- **Colony Failure**: Which party bears the cost of colony fail?.. Colony health and pre-Q Parameters at crop farms must be suitable for colony entry.
- **Arbitration**: Who determines and monitors the rate and type of pesticide usage allowable. Require organic farming certification.

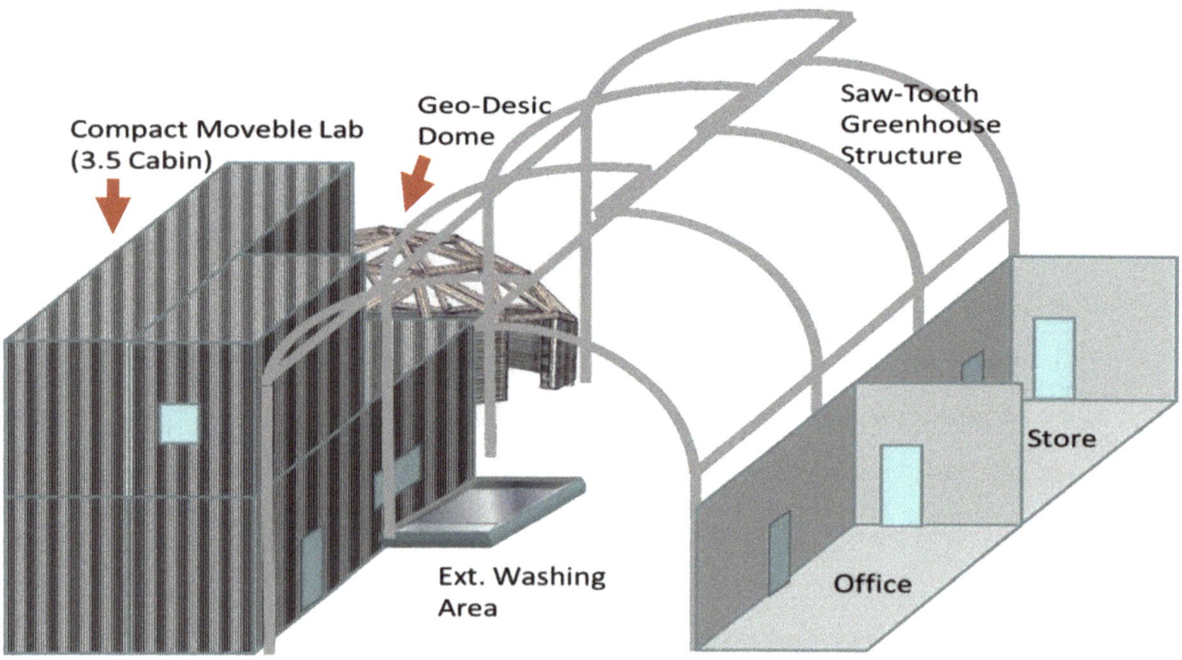

Figure 7 Revamping and redesign of greenhouse features

2.8.3 Portable Cabins and Field Hive Racks

They are constructing specially customized racks for towing box hives to and fro between Farms and Meliponaries.

Figure 8 Portable Hive boxes and Field Hive Racks

Chapter 3

3.0 Crop pollination data on Stingless Bees

List of crops favoured by Stingless bees

Crop 1 Avocado (Persea americana)........ 35
Crop 2 Banana (Musa spp.)........ 36
Crop 3 Brinjals - Eggplant (Solanum melongena)........ 37
Crop 4 Left: Cocoa (Theobroma cacao) flowers, Right: G. thoracica nest entrance at a cocoa tree trunk. 38
Crop 5 Cashew (Anacardium occidentale)........ 39
Crop 6 Chestnut (Castanea sativa, C. mollissima)........ 40
Crop 7 3.7 Chillies - (Capsicum spp.)........ 41
Crop 8 Citrus (Citrus spp.) A. Kaffir Lime, B. Lime, C. Calamansi, D. Kumquat, E.& F. Lemon........ 42
Crop 9 Coconut (Cocos nucifera)........ 43
Crop 10 Coffee (Coffea spp.)........ 44
Crop 11 Cucurbits: Cucurbita (Cucumis spp.)........ 45
Crop 12 Dragon fruit - Pitaya (Hylocereus spp.)........ 46
Crop 13 Durian – Durian Lai (Durio zebethinus, D. kutejensis)........ 47
Crop 14 Grapes (Vitis vinifera)........ 48
Crop 15 Guava (Psidium guajava)........ 49
Crop 16 Langsat (Lansium domesticum)........ 50
Crop 17 Longan (Dimocarpus longan)........ 51
Crop 18 Lychee (Litchi chinensis)........ 52
Crop 19 Jackfruit (Artocarpus heterophyllus) Male inflorescence, Female inflorescence........ 53
Crop 20 Maize (Zea mays)........ 54
Crop 21 Mangosteen (Garcinia mangostana)........ 55
Crop 22 Mango, Kuini, Bacang (Mangifera spp.) Flowers frequently visited by Lepidotrigona sp. 56
Crop 23 Oil palm (Elaeis guineensis)........ 57
Crop 24 Papaya - Pawpaw (Carica papaya) A. Male Papaya B. Female Papaya (inset above - SB foraging sap on papaya leaf)........ 58
Crop 25 Passion Fruit - Markisa (Passiflora edulis)........ 59
Crop 26 Pineapple (Ananas comosus) Shaded Apiary in middle of Pineapple plantation Budding Pineapple Flower 60
Crop 27 Pomegranate (Punica granatum)........ 61
Crop 28 Pomelo & Grapefruit (Citrus spp.)........ 62
Crop 29 Rambutan (Nephelium lappaceum)........ 63
Crop 30 Roselle (Hibiscus sabdariffa)........ 64
Crop 31 Soursop & Annona (Annona muricata, A. squamosa)........ 65
Crop 32 Starfruit - Carambola (Averhhoa carambola)........ 66
Crop 33 Jambos (Syzygium spp.)........ 67
Crop 34 Strawberry (Fragaria spp.)........ 68
Crop 35 Sunflower (Helianthus annuus)........ 69
Crop 36 Snakeskin Fruit - Salak (Salacca spp.)........ 70
Crop 37 Sapodilla – Ciku/Sawo (Manikara zapota)........ 71
Crop 38 Tamarind (Tamarindus indica)........ 72
Crop 39 Tomatoes (Solanum lycopersicum)........ 73
Crop 40 Watermelons (Citrullus lanatus)........ 75

3.1 Avocado (*Persea americana*)

Insects pollinate avocado flowers, so growers bring honey bee hives into the orchard in the flowering season. Most avocado orchards have two different cultivars so that, when some trees have flowers at the female stage, other trees have male flowers. Avocado trees need lots of bees to pollinate all the flowers. Avocados: In general, on a single tree, all the open flowers synchronized. That means they will be all functionally male or all functionally female. The avocado pollen of one tree is compatible with itself and quite capable of pollinating its flowers, known as self-pollination. Do avocados self-pollinate?

In most instances, bees carry the pollen from a type A tree to a type B tree. Avocados are usually not self-pollinated because the male and female parts are not open at the same time. In some climates, avocados may self-pollinate from the wind. http://www.abc.net.au/news/rural/2016-0803/avocado-farmer-swears-by-native-bees-topollinate/7684602

Crop 1 Avocado (Persea americana)

3.2 Banana (*Musa* spp.)

Banana: As the flower stalk grows, it bends downward. The female flowers emerge first, and these are the ones that produce bananas. Sterile and then male flowers form below the female flowers, but these don't develop fruit and usually fall off. Because they don't need pollination, alone "Cavendish" plant will produce fruit.

https://homeguides.sfgate.com/ba nana-trees-self-fruitful-61100.html

Bats pollinate wild bananas and disperse their seeds. But the commercial bananas we eat have been seedless and without the need for pollination for thousands of years. They're grown only from suckers cut off a mother plant and transplanted around the tropics.

Crop 2 Banana (Musa spp.)

3.3 Brinjals - Eggplant (*Solanum melongena*)

Brinjals: Whichever method you use for pollinating eggplants by hand, the ideal time is in the morning between 6 and 11 a.m. However, in a pinch, hand-pollinating eggplants can occur in the afternoon. You'll have success when the flower closes but does not fall from the plant.

Eggplant bears flowers with abundant pollen dispersed through small orifices in the anthers. Moreover, eggplant flowers do not produce floral nectar and bees that pollinate eggplant may be expected to depend on other sources of nectar nearby (Free, 1993).

Eggplant flowers, like tomato flowers, are self-pollinating flowers by their design. They are, however, even easier to get pollinated and have fruit set in the greenhouse than are tomato flowers. To make sure the flowers get pollinated, you can tap the flower stalk with a pencil when the flower petals are fully open. https://www.cropking.com/blog/eggplants-greenhouse

Crop 3 Brinjals - Eggplant (Solanum melongena)

3.4 Cocoa (*Theobroma cacao*)

Crop 4 Left: Cocoa (Theobroma cacao) flowers, Right: G. thoracica nest entrance at a cocoa tree trunk.

Cacao: While many of the world's flowers are pollinated by bees (Hymenoptera) or butterflies/moths (Lepidoptera), cacao flowers are pollinated by tiny flies, forcipomyia midges, in the subfamily Forcipomyiinae[8]. The small chocolate midge is the only pollinator that serves the cacao plant.

Cacao is a small tropical tree that grows in the shade of larger trees in Central America, South America, Africa, and Asia. The trees' flowers grow directly from the trunks and lower branches and, when pollinated, produce the prized seed pods.

Do bees pollinate chocolate?

The chocolate midge does the work. In truth, the chocolate plant, Theobroma cacao, is pollinated by a small fly called a midge. According to the North American Pollinator Protection Campaign, this little fly—only one millimetre long—is the only pollinator that can negotiate its way into the complex chocolate flower. Nov 5, 2014 Chocolate pollination: not the job of bees - Honey Bee Suite https://honeybeesuite.com/chocolate-is-not-pollinated-by-bees/

[8] Editor's note: family Ceratopogonidae, the famous, biting midges of the tropics - a genuine pest to humans

3.5 Cashew Nut (*Anacardium occidentale*)

The pollination requirements of cashew (*Anacardium occidentale* L.), an andromonoecious[9] tree, were studied in NE Brazil, where cashew is indigenous. Cashews: Shown through the use of bagging and caging experiments, controlled hand-pollination and emasculation of flowers that the wind plays little role in cashew pollination and that cashew is self-fertile. However, only

Crop 5 Cashew (Anacardium occidentale)

pollen from the large stamen, either male or hermaphrodite-derived, can affect pollination. **Bees** are the most frequent flower visitors of commercially grown cashew and are implicated as cashew's primary pollinators (Freitas and Paxton, 1996). There is a need to increase raw cashew nut production in India to sustain itself globally. https://www.cambridge.org/core/services/aop-cambridgecore/../S0021859600074876

Insects appear to be the significant cashew pollinators, agreeing with the **pollination** syndrome suggested by its flowers' morphology.

[9] Editor;s note: some flowers that make pollen and others that make both pollen and produce fruit- i.e. are pollinated

3.6 Chestnut (*Castanea sativa, C. mollissima*)

Chestnut: Chestnut is self-compatible but still requires cross-pollination because the male and female flowers do not bloom simultaneously on an individual tree. The flowers are catkins, and a variety of pollinators collect both nectar and pollen from the flowers.

Unlike most other nut trees, the Insects pollinate American chestnuts. In the related Caucasian chestnut tree, *Castanea sativa*, pollination by honeybees can improve total nut yield.

Chestnuts flower about 6 to 8 weeks after growth commences in the spring. They are mainly wind-pollinated and require cross-pollination[10] to set nuts. Therefore, one tree by itself won't have a crop. If you plant seedlings of pure species, pollination is seldom a problem.

Crop 6 Chestnut (Castanea sativa, C. mollissima)

www.pollinator.ca/canpolin/images/PDF_American%20chestnut.pdf

[10] Editor's note thus, are exactly the same as cultivated *Coffea canephora*, the famous robusta coffee grown all over the tropics

3.7 Chillies - (*Capsicum* spp.)

Chillies are pollinated by bees but failing that is also self-pollinating, so a gentle shake of flowers will help ensure that they set[11].

For chillies, bees are not important for pollination.

Crop 7 3.7 Chillies - (Capsicum spp.)

Although you may see plenty of bees in the patch, capsicum is self-pollinated. Bush movement due to the wind is enough for pollination.

Chillies: To hand pollinate your pepper plants, wait until the afternoon (between noon and 3 p.m.) when the pollen is at its peak. Use a tiny artist's paintbrush (or even a cotton swab) to gently transfer the pollen from flower to flower.

You are encouraging fruit. Your chilli plants will produce flowers and then chillies on each of their sides or lateral offshoot[12], so the more side stems you have, the more fruit your plant will grow.

https://homeguides.sfgate.com/pollinate-bell-pepper38916.html

---

Editor's notes

[11]: Some simply will not self-pollinate unless a bee or the wind shakes or vibrates them, especially where humidity is high. The pollen rests INSIDE the anther and is not simply released from them.

[12] It is at the base of the flower, at the ovaries, that fruit grows.

3.8 Citrus (*Citrus* spp.)

Citrus: Do bees pollinate citrus? Citrus flowers are perfect, having both sexes on the same blossom so that self-pollination takes place regardless of pollinators. But bees (pollinators) are distributed throughout citrus groves in any case. Do citrus trees need to be pollinated?

Most citrus trees are grown indoors or outdoors are self-fruitful, including oranges, grapefruit, kumquats, lemons and limes. However, some mandarin orange varieties produce more fruit with cross-pollination, but those fruits may have more seeds than those that self-**pollinate**.

Crop 8 Citrus (Citrus spp.) A. Kaffir Lime, B. Lime, C. Calamansi, D. Kumquat, E.& F. Lemon

Citrus trees require heat, good drainage and moist soil to thrive, but most citrus types, including lemon, are self-fruitful and do not require cross-pollination to set fruit[13]. You can plant one lemon tree in your backyard and enjoy a good fruit yield under acceptable cultivation practices.

https://homeguides.sfgate.com/lemon-treeneed-pollinated-another-lemon-tree-60046.html

[13] Editor's note: A good time to mention another coffee, *Coffea arabica*, which is much the same in that it is self-compatible and transfers pollen to self-flowers, however, being visited by a pollinator like a bee can significantly increase fruit set and mature fruit size Roubik 2002.

3.9 Coconut (*Cocos nucifera*)

Coconut: Coconut palms are primarily cross-pollinated, although some dwarf varieties are self-pollinating. The two main types of coconut palm are tall and dwarf. Tall palms are typically cross-pollinating, while dwarf palms are usually self-pollinating. The coconut plant is monoecious, with inflorescences bearing both male and female flowers.

Crop 9 Coconut (Cocos nucifera)

The palm produces both the female and male flowers on the same inflorescence; thus, the palm is monoecious. Other sources use the term polygamomonoecious[14]. The female flower is much larger than the male flower. Flowering occurs continuously.

https://en.wikipedia.org/wiki/Coconut

[14] Editor's note: in dense plantings, coconut is wind pollinated- as it is on small oceanic islands and strand lines, where there are no flower visiting insects.

3.10 Coffee (*Coffea* spp.)

Coffee: After three to four days, the flowers drop off on the ground, enabling new green beans to emerge from the same bud on the branch of the coffee shrub. Two types of coffee plants cultivated, *Coffea arabica,* is self-pollinating, whereas the Robusta coffee, *Coffea canephora* plant, depends on cross-pollination.

Crop 10 Coffee (Coffea spp.)

After all, the coffee we mostly drink — Arabica — comes from a self-pollinating plant. Still, bees play an essential role in coffee, acting as a kind of pollination booster. Their work means that coffee plants produce 20-25% more fruit.

Three to four years after the coffee is planted[15], sweetly smelling flowers grow in clusters in the axils of the coffee leaves. The fruit is produced only in the new tissue.

https://www.mnn.com/your-home/organic-farming../what-do-bees-have-docoffee-lot

[15] The most commonly cultivated coffee usually produces flowers its first year, and the plant needs to be replaced by six to eight years. This is true for the catimor hybrid grown extensively in Latin America, and elsewhere. Its flavor is inferior and the beans do not sell for nearly as much as the other cultivars of *C. arabica*, whose trees, by the way, are productive for a few decades.

3.11 Cucurbits: Cucurbita (*Cucumis* spp.)

Cucurbits: Cucurbits are monoecious; there are separate male and female blossoms on the same plant. The male flowers tend to open first, followed by the female flowers. It is only when both the male and female flowers are open that pollination can occur. They include cucumber, watermelon, squash, gourd and pumpkin that flower on the same plant. The most successful way for cucurbits to be pollinated is by insects, bees being the most productive pollinators of cucurbits. Many cucumbers don't need insect pollination, despite their large, colourful flowers and the fact that related crops (melons, courgettes, squashes, etc.) generally require pollinators[16]. Indeed, some varieties taste bitter if they pollinate. This well-pollinated cucumber plant is growing a new cucumber. The fruit comes from the female flower, pollinated by pollen transferred from a male flower, like the bright yellow one above. Ideally, pollen is transferred by bees, but if not, hand-pollination is an option. https://bonnieplants.com/2012/07/pollination-problems-give-handpollination-a-try/

Crop 11 Cucurbits: Cucurbita (Cucumis spp.)

[16] Editor's note: The 'squash and gourd bees', family Apidae, are a couple solitary genera that are specialized to gather the large pollen grains, and the adults live in the flowers, resting there at night. These are far and away the major pollinators- and they are Neotropical. In the rest of the tropics a variety of bees visit the flowers, many of which do not pollinate the plant.

3.12 Dragon fruit - Pitaya (*Hylocereus* spp.)

Dragon fruit: Also known as pitaya, the dragon fruit plant is native to Mexico and South and Central America. Approximately 25 species of dragon fruit exist, each producing oversized, showy flowers. Because the flowers are night

Crop 12 Dragon fruit - Pitaya (Hylocereus spp.)

bloomers and not capable of self-pollination, they are highly dependent on bats and moths to do the job. A dragon fruit plant requires a second dragon fruit plant's pollen -- of a different variety[17] -- to set fruit. In place of bats and moths, hand pollination (manual) is necessary.

Hylocereus undatus "Vietnamese giant."

This variety can self-pollinate, which means it can produce fruit on its own without help from you. Be aware that other types may require cross-pollination with another plant. https://www.miamiherald.com/living/ho me-garden/article1977019.html

[17] Editor's note: It is a CACTUS. And here, just like a lot of commercial apples, Malus, the pollen from a compatible cultivar is required for fruit set. They cannot pollinate other clones of their kinds.

3.13 Durian – Durian Lai (*Durio zebethinus, D. kutejensis*)

Durian: These fruit bats, together with nectar-eating bats, are also the creatures responsible for the pollination of tropical and subtropical trees and shrubs. And where the durian tree is concerned, it is the Dawn bat or Cave Nectar Bat (*Eonycteris spelaea*) that is the only effective pollinator of its flower[18].

Crop 13 Durian – Durian Lai (Durio zebethinus, D. kutejensis)

Even if the female and male flower parts of durian flowers were active at the same time, most durian trees have a high degree of self-incompatibility. In other words, the flowers cross-pollinate from different trees to set fruit.

Durian (*Durio* spp.) is a fruit-bearing tree in the Hibiscus family, Malvaceae. A 2008 study of the pollination ecology of Durian in Southern Thailand found that while honeybees were the most frequent visitors to Durian flowers, only fruit bats acted as pollinators. June 14, 2017, https://www.quora.com/How-do-Ipollinate-durian-trees-in-a-zone-with-nofruit-bats

[18] Editor's note: flowers are intensely visited by the giant honeybee, *Apis dorsata*, but rarely pollinated by them.

3.14 Grapes (*Vitis vinifera*)

Crop 14 Grapes (Vitis vinifera)

Grapes: Most cultivated *Vitis vinifera* grapevines are hermaphroditic, with male stamens and female ovaries, while many wild grapes[19] are either male, producing pollen but no fruit, or female, producing fruit only if a pollinator is nearby.

The grapes do not need bees to pollinate - they are wind-pollinated. We rarely see honeybees in the vineyard during bloom. Most of the time, like now, we do not even have honeybees on our property - they are at ten different locations (farms) in the surrounding counties. We do not grow completely organic grapes.

The three main varieties of grapes include muscadine, European and American. Muscadine grapevines are either male or female and need cross-pollination to mature into fruit. European and American have both male and female parts on each flower and self-pollinate. https://homeguides.sfgate.com/crosspollinate-grape-vines-24083.html

[19] Editor's note: Commercial grapes are produced by application of plant growth hormones to flowers, which produce uniform, large, seedless grapes.

3.15 Guava (*Psidium guajava*)

Crop 15 Guava (Psidium guajava)

Guavas are primarily self-fruitful, although some strains seem to produce more fruit when cross-pollinated with another variety. Guavas can bloom throughout the year in mild-winter areas, but the most massive bloom occurs with the onset of warm weather in the spring.

Many studies have shown that guava is a highly cross-pollinated crop, and pollination by insects, particularly honeybees, results in higher fruit yield. The flowers of guava are cyme or single. The receptacle of guava is hollow and united to the ovary.

ROLE OF HONEY BEES IN THE POLLINATION OF GUAVA (Psidium ..

krishikosh.egranth.ac.in/bitstream/1/5810024175/1/Th9977.pdf

3.16 Langsat (*Lansium domesticum*)

Native to Southeast Asia, *Lansium domesticum* is known by various common names such as Langsat, Lansa, Lanzon or Lanzones. It is a tropical tree of about 30 m in height and 75 cm in diameter. The bark is grey with light and dark spots. The hairy leaves are pinnately compound. Flowers are small with short stems and often hang from large branches or the trunk. The fruits can be elliptical, oval, or round, pale yellow when ripe, and forms long drooping clusters of 25 fruits per cluster.

Crop 16 Langsat (Lansium domesticum)

When ripe, fruits develop brown scurf on their thin skin. It is eaten fresh, made into candies, preserved in syrup, and processed into wine. The resin is non-toxic and used against diarrhoea and intestinal spasms. Dried peel burned to repel mosquitoes and as incense. The plant is not tolerant of drought, is slow-growing and shallow-rooted. It is grown by seeds or by budding.

Insects pollinate the flowers. Suitable for: medium (loamy) and heavy (clay) soils and prefers well-drained soil.

https://pfaf.org/user/Plant.aspx?LatinName=Lansium%20domesticum

3.17 Longan (*Dimocarpus longan*)

Crop 17 Longan (Dimocarpus longan)

Longan is a cross-pollination plant. The ratio between male and female flowers is 6:1, and the blossom of both kinds of flowers do not synchronize. The rate of fruit set of longan was significantly higher than longan without insect pollinators. Yields of 6 years longan pollinated by honeybee open field condition and in a cage were about 30 and 12 times of longan without insect pollinators. The actual yield of longan was as follows, an open field with pollinators was 20+-17.3 kgs/tree, in a cage with bees was 8+-6.5 kgs/tree and in a cell without bees was 0.6+-1.2 kgs/tree.

Yields of 9 years longan pollinated by honeybee open field condition and in a cage were about nine times of longan without insect pollinators. The longan's actual result was as follows: open field pollination was 63.6+-30.0 kgs/tree, in a cage with bees was 61.6+-34.0 kgs/tree and in a cell without bees was 6.8+-3.2 kgs/tree. agris.fao.org/agris-search/search.do?recordID=TH2000002595

3.18 Lychee (*Litchi chinensis*)

Crop 18 Lychee (Litchi chinensis)

Lychees: Self-pollination may occur in lychee through wind pollination; however, flowers are generally recognized as self-sterile[20] , and insects are beneficial because they can transport pollen from anther to stigma for the fruit to set.

Flower Pollination: There must be an overlap of the stages to assure good pollination. On a single tree, panicles emerge and develop at different times, ensuring that there will be pollen available to fertilize female flowers and obtain a good fruit set, assuming that there is a pollinator like bees to carry the ripe pollen from the male to the female flowers. Having multiple trees, such as in a grove situation, or having a tree of a different cultivar with different stages of panicle development assures that there will be an improved overlap of pollen-bearing male flowers with receptive female flowers.

To a small degree, you can get some minimal wind pollination, but insects such as bees, wasps, beetles and flies do most of the work.

Lychee Flower Pollination - www.lycheesonline.com/FlowerPollination.cfm

[20] Editor's note: the usual botanical or pollination term is 'self-incompatible', but self-sterile seems to mean the same thing here. If it is hermaphroditic but has the pollen in the wrong place to reach the stigma it can be both, however (it is self-compatible, but it is sterile- having no way to produce fruit). And a male flower is also, by definition, sterile (no fruit) but it can be successful in pollination.

3.19 Jackfruit (*Artocarpus heterophyllus*) & Breadfruit (*Artocarpus altilis*)[21]

Jackfruit: cultivars set fruit with self-pollination. Seed set with fruit characters such as size, flesh percentage and edible percentage greatly enhanced with cross-pollination.

Jackfruit Tree Pollination. Jackfruit Trees are self-fertile. You will get fruit with only one plant. However, adding a Jackfruit Tree will drastically increase the size of your crop.

Crop 19 Jackfruit (Artocarpus heterophyllus) Male inflorescence, Female inflorescence

https://www.fast-growing-trees.com/Jackfruit-Tree.htm

Breadfruit Pollination: Birds, bats and insects pollinate wild types. Flowers produce nectar—male and female flowers on each tree. The seedless variety does not require pollination.

Ref.: POLLINATION REQUIREMENTS OF SOME TROPICAL CROPS

rfcarchives.org.au/Next/CaringForTrees/Pollination1-88.htm

[21] Editor's note: these are cauliforous, arising directly from the trunk. Jackfruit wood is also highly prized as lumber.

3.20 Maize (*Zea mays*)

Maize (called corn in some parts of the world) is pollinated by wind. The male anthers let go of their pollen, and it blows over to a nearby female flower on another corn plant. Most of the flowers are either male or female on a corn plant (monoecious) rather than both sexes in one flower (hermaphrodite).

Under self-pollination, the silks of an ear are pollinated by pollen from the same plant. Typically, little self-pollination occurs in a field of corn. Most silks of a given plant pollinated by pollen from surrounding plants are known as cross-pollination

Other bees tend to favour one species at a time, therefore do most of the actual pollination. The essential staple food crops on the planet, like corn, wheat, rice, soybeans[22] and sorghum, need no insect help; they are wind-pollinated or self-pollinating.

Crop 20 Maize (Zea mays)

passel.unl.edu/pages/informationmodule.php?idinformationmodule=1075412493

https://en.wikipedia.org/wiki/List_of_crop_plants_pollinated_by_bees

[22] Editor's note: Soy flowers contain nectar, which is not the case for wind-pollinated plants. SOY IS enhanced in fruit and seed set by bees. Universally. The agronomists were slow to see this, but it is well-established. It is like the coffee arabica story- the technicians 'knew' that it was self-compatible and could self-pollinate, so therefore (to them) it had no benefit from bees or other pollinators, which was totally WRONG.

3.21 Mangosteen (*Garcinia mangostana*)

Mangosteen trees were initially thought to be all genetically identical. Now known this is not true, there are small genetic differences. However, pollination is not required to set fruit.

Crop 21 Mangosteen (Garcinia mangostana)

Male and female flowers develop on separate trees (dioecious plants). Mangosteen is an apomictic plant, which means that female trees produce fruit without pollination.

www.softschools.com/facts/plants/mangosteen_facts/1843/

3.22 Mango, Kuini, Bacang (*Mangifera* spp.)

Mango: trees **flower** in synchrony and bear many **flowers blooming** branches all over the tree. The **mango flowers** are simple, open and small in size. They are visited and **pollinated** by many different kinds of insects. **Mango pollinators** include various types of wasps, flies, stingless bees and even ants. The **mango flowers** are simple, open and small in size. They are visited and **pollinated** by many different kinds of insects.

Crop 22 Mango, Kuini, Bacang (Mangifera spp.) Flowers frequently visited by Lepidotrigona sp.

As they visit each flower, their bodies become covered in pollen. Their movements on the flowers also enhance self-pollination on the perfect flowers -- mango trees using both cross- and self-pollination techniques for successful fruiting.

Ref: How Pollinators Make Mangoes! – Discover Pollinators - discoverpollinators.org/how-pollinators-make-mangoes/

3.23 Oil palm (*Elaeis guineensis*)

Oil Palm: Each tree bears male and female **flowers** on separate inflorescences, requiring pollen to be transferred by wind or insects. In the beginning, **oil palm** plantations in Malaysia relied mainly on wind **pollination**. Found to have the highest capacity for pollen cartage of all **pollinators** (Syed, 1979).

Crop 23 Oil palm (Elaeis guineensis)

The bulk **of palm** species are, however, **pollinated** by insects and, in particular, beetles, **bees** and flies. Insect-mediated wind **pollination** proposed for one species whereby minute thrips and beetles mobilize pollen that is subsequently carried to the female flowers by wind

http://eprints.utar.edu.my/1987/1/Effectiveness_of_the_oil_palm_pollinating_weevil,_Elaeidobius_kamerunicus,_in_Malaysia_-_T.M._Teo.pdf

3.24 Papaya - Pawpaw (*Carica papaya*)

The bulk **of palm** species are, however, **pollinated** by insects and, in particular, beetles, **bees** and flies. Insect-mediated wind **pollination** proposed for one species whereby minute thrips and beetles

Crop 24 Papaya - Pawpaw (Carica papaya) A. Male Papaya B. Female Papaya (inset above - SB foraging sap on papaya leaf)

mobilize pollen that is subsequently carried to the female flowers by wind

Flowers on female papaya trees grow in small clusters or as solitary blooms that are larger than male flowers but on shorter stalks. They do not possess stamens and can produce fruit if pollinated. However, like male papayas, they are susceptible to changing gender.

https://homeguides.sfgate.co m/male-female-papaya-trees-56987.html

3.25 Passion Fruit - Markisa (*Passiflora edulis*)

Passion Fruit[23]**:** If using nail clippers, snip the stamen from inside the **flower**. Then just transfer the pollen to the female organ, the pistil, by gently rubbing the brush or swab against it. **Passionflowers** have three pistils. That's all there is to hand **pollination of passion** vines.

Crop 25 Passion Fruit - Markisa (Passiflora edulis)

Gather pollen for hand-pollination within four to six hours after the passion fruit vine flower opens in the morning. The flowers contain both male and female parts but are self-sterile, meaning the pollen must be gathered from one flower and transferred to a flower on a different passion fruit vine.

[23] Editor's Note: Native to tropical America, its pollinators are the large carpenter bees, Xylocopa. They also visit the flowers, as do honeybees, in Africa and Asia. Both are pollinators, but the larger bees do a much better job.

3.26 Pineapple (*Ananas comosus*)

Pineapple Pollination. In the wild, **pineapples** are **pollinated** primarily by hummingbirds. Individual wild **pineapples** are foraged and **pollinated** at night by bats. This is under cultivated[24]because seed development diminishes **fruit** quality, **pollination** is performed by hand, and seeds are retained only for

Crop 26 Pineapple (Ananas comosus) Shaded Apiary in middle of Pineapple plantation Budding Pineapple Flower

breeding.

Reproduction. That is because commercial **pineapple** plants reproduced as not to have seeds. After all, they are said to worsen the quality of the fruit. Wind **pollination does** not occur in **pineapples** because the pollen is too sticky. It is known to be **pollinated** by hummingbirds, honeybees, and **pineapple** beetles.

If the flowers are to produce seeds, they must be fertilized by a **pollinator** (usually a hummingbird) because **pineapples** are incapable of **self**-fertilization https://creationwiki.org/Pineapple

[24] Editor's Note: I think all the commercial growers use parthenocarpic strains- they certainly do in Latin America. It is like banana pollination- it can happen, but it is not desirable or needed.

3.27 Pomegranate (*Punica granatum*)

Pomegranate: *Punica granatum* - Hand-**pollinating pomegranates** requires a delicate, sable artist paintbrush (or even a cotton swab) and a light hand. Gently transfer the pollen from the male stamen to the female ovary. If you have multiple trees, go from tree to tree to cross-**pollinate**, which will increase the crop.

Crop 27 Pomegranate (Punica granatum)

Most **pomegranates** are self-fruitful, meaning they **do** not require another **tree** to cross-**pollinate** with, as the **bees do** all the work. That said, **planting** another **pomegranate** nearby can increase fruit production on both plants. A little cross-**pollination** can't hurt, but it isn't necessary.

https://www.gardeningknowhow.com/ ../pomegranate/pomegranate-tree pollination.htm

3.28 Pomelo & Grapefruit (*Citrus* spp.)

Pomelo: Citrus maxima is Self-**Pollinating**. The pomelo fruit is the largest of all citrus fruits. It has a very thick rind (much thicker than in any other citrus) and is filled with sweetish yellow or pink flesh. http://world-crops.com/pomelo/

Crop 28 Pomelo & Grapefruit (Citrus spp.)

Flowers: Large flowers of 3-7 cm in diameter, either single or in small clusters, with cream-coloured petals.

Leaves: Pomelo has the largest leaves among all citrus.

Grapefruit flowers are self-pollinating. The pollen from the tree's flowers pollinates the other flowers: there is no need for a second tree. The grapefruit tree depends on bees and other insects to spread the pollen from blossom to blossom.

3.29 Rambutan (*Nephelium lappaceum*)

Rambutan: Aromatic rambutan flowers are beautiful to many insects, especially bees. Flies (Diptera), bees (Hymenoptera), and ants (*Solenopsis*) are the primary pollinators. Among the Diptera, *Lucilia* spp. is abundant, and among the Hymenoptera, honeybees (*Apis dorsata* and *A. cerana*) and the

Crop 29 Rambutan (Nephelium lappaceum)

stingless bee are the significant visitors[25].

Apis cerana colonies foraging on rambutan flowers produce large quantities of honey. Bees foraging for nectar routinely contact the stigma of male flowers and gather significant amounts of the sticky pollen from male blossoms. Little pollen has been seen on bees foraging female flowers. Although male flowers open at 06:00, foraging by *A. cerana* is most intense between 07:00 and 11:00, tapering off rather abruptly after that. In Thailand, *A. cerana* is the preferred species for small-scale pollination of rambutan. Its hair is also helpful in pollination where pollen can be hooked on and transported to female flowers.. https://en.wikipedia.org/wiki/Rambutan

[25] Editor's note: Our work identified *Scaptotrigona* in Mexico as the main pollinators, which is also true for avocado there.

3.30 Roselle (*Hibiscus sabdariffa*)

Roselle: Flowers of **roselle** are short-lived, remaining open for not more than three hours, closing at about noon of the same day they open. **Roselle is** almost entirely self-**pollinating**, though some cross-**pollination** by insects may occur.

Crop 30 Roselle (Hibiscus sabdariffa)

Two local varieties of roselle (*Hibiscus sabdariffa* L.), Jamaican Green and Jamaican Red, were grown to determine the amount of natural cross-pollination. Roselle is an essential crop in Jamaica, where traditionally, the fleshy calyces are used for the preparation of a drink during the Christmas season.

In roselle, *Hibiscus sabdariffa* L. (Malvaceae) natural cross-pollination occurs. www.scielo.br/scielo.php?script=sci_arttext&pid= S1415-47572000000300027

3.31 Soursop & Annona (*Annona muricata, A. squamosa*)

Soursop: The **Soursop** Tree is **self-pollinating**; however, if it has a mate to cross-**pollinate** with, it **will** produce a higher fruit yield. You **do** not have to buy another plant.

Pollinators transfer pollen and seeds from one flower to another, fertilizing the plant so it can grow and produce food. Cross-pollination helps at least **30 %** of the world's crops and **90 %** of our wild plants to thrive.2 Without bees to spread seeds, many plants—including food crops—would die off.

Although the **Soursop Tree** is self-**pollinating**, however, if it has a mate to cross-**pollinate** with, it will produce a higher fruit yield. https://www.fast-growing-trees.com/Soursop-Tree.htm

Crop 31 Soursop & Annona (Annona muricata, A. squamosa)

Pollination: Dynastid scarab beetles appear necessary within the genus *Annona*. Those species of *Annona,* which are more morphologically derived, as well as all *Rollinia* spp., possess reduced floral chambers and attract small beetles.

3.32 Starfruit - Carambola (*Averrhoa carambola*)

Starfruit: **Pollination**. **Flowers** on a **carambola** variety have either long or short styles, a

Crop 32 Starfruit - Carambola (Averhhoa carambola)

condition called heterostyly. Buds may remain on the tree for up to 16 days even when they are no longer receptive. Some **carambola** growers keep a few hives of honey **bees** in the orchard for **pollination**.

www.itfnet.org/gfruit/Templates%20English/carambola.flower.fruit.htm

Some **carambola** cultivars may require cross-**pollination** (short-styled by long-styled cultivar or vice versa) for good **fruit** set and yields. **Carambola** growers must ensure that there are enough insect **pollinators** to **pollinate** the flowers.

3.33 Rose Apple, Jambu (*Syzygium jambos*)

Syzygium: Natural pollen transfer in the species was efficient, and fruit set following open pollination was quite high. It is a crop that is most suited to insect and wind pollination. During flowering periods, they observe the bees' foraging behaviour and underscore certain plant species' importance in the brood's feeding. The nectar collected in particular flowers, such as pomegranate, or the sap from some leaves, such as the fig tree, has unique properties.

Crop 33 Jambos (Syzygium spp.)

https://www.researchgate.net/publication/289979033_Reproductive_phenology_flower_biology_and_pollination_in_jamun_Syzygium_cuminii_L [accessed Sep 29 2018].

Cross-pollination helps at least **30 %** of the world's crops and **90 %** of our wild plants to thrive[26]. Several of these names are also applied to other *Syzygium* species, while "jambu" can also mean guava in Malay.

[26] Editor's note: genus Syzygium is heavily visited by butterflies and moths- right in my yard in Panama.

3.34 Strawberry (*Fragaria* spp.)

Strawberries: Strawberries are self-fertile. Strawberry plants will pollinate themselves, but they usually need the assistance of wind or pollinators, such as bees, to do the work of transferring the pollen from the stamens, the male parts of the flowers, to the stigma, the female part of the flower.

Yield of strawberry cultivars pollinized by jatai bees under protected environment by Odirce Teixeira Antunes et al. Escola Agrotécnica Federal de Sertão, C. Postal 21, 99170-000 Sertão-RS; [3]Fac. de Agronomia e Medicina Veterinária (FAMV) da UPF, C. Postal 611, 99001-070 Passo Fundo-RS;

Crop 34 Strawberry (Fragaria spp.)

Pollination of Strawberry in Kenya, by Stingless Bees (Hymenoptera: Meliponini) and Honeybee (Hymenoptera: Apini) for Improved Fruit Quality By Asiko Grace Adala, B.Sc, University of Nairobi, M.Sc. Utrecht University, Netherlands.

Kakutani, Takehiko & Inoue, Tamiji & Tezuka, Toshiyuki & Maeta, (Yasuo, 1993)[27]. Pollination of strawberry by the stingless bee, *Tetragonula minangkabau*, and the honey bee, *Apis mellifera*: An experimental study of fertilization efficiency. Researches on Population Ecology. 35. 95-111. 10.1007/BF02515648.

[27] Editor's note: These friends from Kyoto are T. Inoue, Y. Maeta, T. Tezuka and T. Kakutani.

3.35 Sunflower (*Helianthus annuus*)

Sunflower seed: The showy outer ray petals help attract the pollinators. Bees go from flower to flower within the disc, becoming covered with pollen. They then pollinate other sunflowers as they go from plant to plant. The success of sunflower as a crop for seeds and oil depends totally on bees. Do Bees Pollinate Sunflowers? | Home Guides | SF Gate https://homeguides.sfgate.com/bees-pollinate-sunflowers-65431.html Are sunflowers self-pollinating?

Crop 35 Sunflower (Helianthus annuus)

The pollen/sperm can belong to the original plant or may come from another **sunflower**. Self-pollinating or Self-pollination is a mechanism that this flower uses to reproduce. On the occasion where the stigma receives no pollen, the stigma will twist and can move to contact anthers,

3.36 Snakeskin Fruit - Salak (*Salacca* spp.)

Salacca: Pollination. Salak plants are either female or male and pollinated by several native pollinators, including stingless bees. But these pollinators are inefficient, with only 10 to 15 % of flowers producing fruit[28]. By hand-pollinating Salak, you can quickly achieve close to 100% pollination.

Flower: Dioecious; requires male and female plants as are necessary for fruiting

Fruit: Rich yellow-white meat is slightly crisp, a delicate, delicious blend of acidity and sugars.

Crop 36 Snakeskin Fruit - Salak (Salacca spp.)

Pollination: Usually dioecious, and some are monoecious. The fruits mature in 5-7 months after pollination.

A monoecious variety, 'Bali' produces inflorescences with both hermaphrodite and staminate flowers, i.e., it is andromonoecious. The latter produce functional pollen.

Season: In principle, these palm flowers and fruits continuously throughout the years but still have peak periods during the year

https://www.growables.org/information/TropicalFruit/PalmSalak.htm

[28] Editor's note: For chocolate, it is something like 1%. Inefficiency and efficiency are in the mind of growers. Too many fruits, and the tree dies the next year- this happens in cacao and peaches, for example, and to some extent among ALL perennials.

3.37 Sapodilla – Ciku/Sawo (*Manikara zapota*)

Sapodilla (Ciku/sawo): In self-incompatible cultivars, the flowers require cross-pollination by another sapodilla plant or variety to produce fruit. Other types may not require cross-pollination but produce more fruit when cross-pollinated.

The thrips take shelter in the flowers and live on the pollen grains, nectar, and stigmatic exudations. Bees have been observed in flowering

Crop 37 Sapodilla – Ciku/Sawo (Manikara zapota)

sapodilla plantings (Sambamurty and Ramalingam 1954), although examining the insects indicated that they were not carrying pollen. https://hort.purdue.edu/newcrop/proceedings1996/V3-439.html

3.38 Tamarind (*Tamarindus indica*)

Tamarind: *Apus mellifera* is also the only species that visited the flowers throughout the day. For self-incompatible (allogamous) plants with low tree densities (like tamarind), *A. mellifera* is not an efficient pollinator.

Crop 38 Tamarind (Tamarindus indica)

www.academicjournals.org/article/article1419951831_DIALLO%20et%20al.pdf

Sex **distribution:**

Bisexual (each flower of each individual has both male and female structures). ·\

Mode **of pollination:** Tamarind a predominantly cross-pollinated. The flowers are nectariferous. Pollination is through the agents of insects. In India, a red ant (*Oecophylla smaragdina* Fab.) plays an important role.

Apis mellifera and Stingless Bees are listed as visitors and pollinators in the Sudanian Region of Burkino Faso (Diallo et al. 2014)

3.39 Tomatoes (*Solanum lycopersicum*)

Crop 39 Tomatoes (Solanum lycopersicum)

Cross-section (top photo) and longitudinal (bottom photo) of the tomato. Fruits on the left are fruits produced in flowers not bagged since the flowers on the right are fruits of flowers are bagged. Statistical tests demonstrate the difference between production in the two study groups. The number of seeds, the size and the weight of the tomatoes are higher in the fruits not bagged

Tomatoes:

Do tomato flowers need pollination? Many edible plants, such as tomatoes, are self-fertilizing or self-pollinating. You only need one plant in the garden for the fruit to set. The male and female parts are contained in the same flower. It truly makes tomato plants a top contender for container gardening.

https://www.gardeningknowhow.com/edi ble/vegetables/tomato/pollinatetomatoes.htm

Commercial tomato flowers are buzz pollinated[29], and typically by bumblebees, the lack of or low insect numbers can inhibit the pollination process.

In these situations, you may need to hand pollinate tomatoes to ensure improved tomato plants' yield.

[29] Editor's note: There is no wind pollination of buzz pollinated flowers. They require bees that sonicate or vibrate the flowers- and as far as I know, no Asian stingless bee does this, nor does any honeybee...

3.40 Watermelons (*Citrullus lanatus*)

Watermelon in nature is pollinated by bees, wasps and other nectar seeking insects who hop from flower to flower and distribute pollen. Watermelon farms rely on bees to pollinate the crop, and farmers will hire and raise bees to perform this service. In order to pollinate the watermelon plant, pollen from the male flowers must make its way to the female flowers by one method or another.

https://homeguides.sfgate.com/watermelon-pollinating-procedures-24371.html

Crop 40 Watermelons (Citrullus lanatus)

Honeybees are considered the most effective pollinating insect for melon crops. Cross-pollination produces slightly heavier fruit. Watermelon is self-fertile, with the female flower being pollinated equally well by pollen from a male flower on the same or a different plant[30].

[30] Editor's note: cross pollinated- the pollen must move to a female flower from a male one. Little bees that visit a bunch of male flowers are not accomplishing any pollination. This is not obvious to everyone,

#	Crop Visited by SB	Genus	Self-Pollination	Normal Pollinating agent	Meliponiculture Resource	Tree Usage (Host) & Fruit Forage	Remarks (1 -20)
1	Avocado	*Persea*	No	Bees & wind*	Pollen & Nectar	Trunk	Overipe flesh food source
2	Banana	*Musa*	Yes	bats on wild var.	Nectar & Pollen	Smoked bananas	Cavendish cultivars & Plantains
3	Brinjals	*Solanum*	Yes	Bees	Pollen		Eggplants & aubergines
4	Cocoa	*Theobroma*	No	Tiny flies (midge)	Rind Sap & Pollen	Fruit Rind	Resin
5	Cashew	*Anacardium*	Herma*	Bees	Nectar & Pollen	Trunk	Possibly Toxic Resin
6	Chestnut	*Castanea*	Cross	Bees & wind*	Nectar & Pollen	Trunk & Buttress	Improvr nut yield with bees
7	Chillies	*Capsicum*	Yes	Bees & wind*	Nectar & Pollen		Inc Paprika Peppers & Jalapeno
8	Citrus	*Citrus*	Yes	Bees & wind*	Nectar & Pollen	Trunk	Limes, Calamansi, Tangerines, Lemons
9	Coconut	*Cocos*	Self & Cross	Insects	Nectar & pollen	Trunk & Kernel Host	Dwarf var self poll
10	Coffee	*Coffea*	Self & cross	Bees as booster	Nectar & Pollen	Ground coffee	Arabica self; Robusta cross
11	Cucurbits	*Cucumis, Cucurbita*	No	Bees	Nectar & Pollen		Cucumber, Pumpkins, Gourds, Melons
12	Dragonfruit	*Hylocereus*	No	Bats & Moths	Nectar & Pollen		Bee visit early morn
13	Durian	*Durio*	No	Bats	Nectar	Trunk	Inc Durian Lai (red flesh)
14	Grapes	*Vitis*	Herma*	Wind	Rare		Rare visits
15	Guava	*Psidium*	Yes	Bees & wind*	Nectar & Pollen	Trunk	More fruits when cross by bees
16	Jackfruit	*Artocarpus*	Yes	Self fruit	Gum	Trunk & Fruit	Size & flesh % enhanced by cross
17	Langsat	*Lansium*	No	Insects	Nectar & Pollen	Trunk	Overipe flesh food source
18	Longans	*Dimocarpus*	Cross 6:1	Bees	Nectar & Pollen	Trunk	Higher fruit set ~9x
19	Lychees	*Litchi*	Yes	Wind & insects	Nectar from calyx	Trunk	Overipe flesh food source
20	Maize	*Zea*	Yes	Wind	Pollen		Corn & cereals

Table 1 Top 40 cash crops visited by Stingless bees in Malaysia (1 -20)

#	Crop Visited by SB	Genus	Self-Pollination	Normal Pollinating agent	Meliponiculture Resource	Tree Usage (Host) & Fruit Forage	Remarks (21-40)
21	Mangosteen	*Garcinia*	Yes	Self fruit - f trees	Nectar & resin	Trunk & Trunk resin	Kuini, Bachang & Apple Mango
22	Mango	*Mangifera*	Cross & Self	Insects	Nectar & Pollen	Trunk & Rind Sap	Lepidotrigona spp
23	Oil Palm	*Elaeis*	Yes	Wind & insects	Pollen	Rotted Trunk	Allergenic pollen*
24	Papaya	*Carica*	Herma*	Self fruit	Gum, Nectar & ♂Pollen	Leaf & Rind sap	Pawpaw
25	Passion Fruit	*Passiflora*	Self Fertile	Bees, wasps & Bats	Nectar & Pollen	Leaf phloem	Xylocopa & Some self pollinate
26	Pineapple	*Ananas*	No	Self fruit, Hummingbirds, bees & beetle	Nectar & sticky pollen	Fruit peel	may produce seeds when cross
27	Pomegranate	*Punica*	Yes	self fruitful	nectar & Pollen		multiple trees to increase crop
28	Pomelo	*Citrus*	Yes	Bees help	Nectar & Pollen		Inc. Grapefruit
29	Rambutan	*Nephelium*	No	Bees	Nectar	Trunk & Branch	Overipe flesh feed
30	Roselle	*Hibiscus*	Yes	Insects sometimes	Nectar from calyx	Calyx sap	
31	Starfruit	*Averrhoa*	No - cross	Insects	Nectar & Pollen		Overipe flesh feed
32	Soursop	*Annona*	Yes	Insects	Nectar & Pollen	Rind sap	Higher fruit yeild on cross
33	Syzygium	*Szygium*	Open	Selfing & wind	Pollen & Nectar		Overipe flesh feed
34	Strawberries	*Fragaria*	Yes	Wind	Pollen &/or Nectar		
35	Sunflower	*Helianthus*	Yes	Selfing & Bees	Pollen & Nectar		Confectionery hybrids
36	Sapodilla	*Manikara*	Cross	Insects	Pollen & rind sap	Trunk & Sap	Produce more fruit when cross
37	Salacca	*Salacca*	No	Bees	Pollen & Nectar		SB -Only10-15% of flowers fruit
38	Tamarind	*Tamrindus*	Cross	Insects	Nectar	Trunk, Shoot phloem	HB not efficient
39	Tomatoes	*Solanum*	Yes	Wind & Bees	Pollen		No air movements require bees.
40	Watermelon	*Citrullus*	No	Bees	Pollen & Nectar		Heavier fruit when cross

Table 2 Top 40 cash crops visited b y Stingless bees in Malaysia (21-40)

3.41 More Research Needed

- **Hermaphroditic** flowering cash crops like papayas & grapes and nocturnal blooming like Dragon fruit & Durians get little influence in the pollinating yields when visited by Meliponines purely because the crops are self-fruiting and being nocturnal blooming[31]. (about 10%)

- **Assisted Pollination**: There are indications that the following tropical crops benefit from insect pollination: Avocado, Cashew, Coconut, Coffee, Guava, Lychee, Mango, Melon, Passion Fruit, Safflower, Sunflower, and many varieties of Citrus and Cucurbits. (24/40 [60%] incidentally or accidentally pollinated by bees)

- **Increased Yield**: Likely, some (for example, Mangosteen, Pomegranate, Jackfruit, Roselle) give only moderately increased yield when pollinated by insects while others (for instance, Longan and Guava) may result in some better gains. (Approx. 15%)

However, more research is needed. Other crops' pollination requirements, including tropical legumes used in large-scale production, are mostly unknown. Because of the lack of time and facilities, many studies on crop pollination have been preliminary only. Careful studies instigated to determine:

a) whether fruit or seed set is satisfactory;
b) whether the crops benefit from pollination by an increase in yields, fruit size, flesh mass
c) which primary agents do pollinate the flowers[32].
d) Which crops are readily robbed[33] of nectar and pollen by bees without ever pollinating the crop?

Editor's notes:

[31] if they flower at night, like many of the flowers taken by Neotropical bees, they are visited for nectar and pollen very heavily the next morning. However, the stigmas are no longer receptive, so no pollination can occur

[32] f or an apple or pear orchard, 5-8% is enough, but for seed crops on annuals, growers like to see 85% or more. It really depends on what it is and how costly or valuable each individual plant is to maintain and harvest, plus the market value. A huge number of ugly apples versus fewer beautiful ones are all a grower needs to cosndier, if there is a market for the beautiful ones (and there is). Apples in USA, Japan, S. Africa and elsewhere are a luxury item- grown for beauty and other qualities. The ones they produce in India go to make cider.

[33] not called robbing, when it is inefficiency on the part of the bee-flower interaction, which is certainly to be expected with crops, many of which are neither native nor natural varieties...

Chapter 4

4.0 Meliponiculture of *Geniotrigona thoracica* (Smith 1857) in Malaysia

4.1 Abstract

By Abu Hassan Jalil[1] (abuhsn@gmail.com) & Hendra Ja[1]

[1] Akademi Kelulut Malaysia Sdn Bhd

Meliponiculture, like the more established Apiculture, deals with Meliponines rather than Apis bees. Rearing Meliponines pose many different variations because their nest structures are different and require different methods of culture. Harvesting procedures are different for the honeypots of varied sizes, and their propolis textures also differ and require different approaches to the development of value-added derivatives products. There are over 80 valid species in the Indomalaya-Australasian region (Rasmussen, Thomas & Engel 2017). Of these, *Geniotrigona thoracica* has the largest workers. This paper will address issues that pertain to initiating and acquiring colonies, providing foraging sources and meeting challenges in culturing and harvesting stingless bee products. We look at the variations and morphs of *Geniotrigona* Moure, 1961 and its two species' taxonomic status. The caste and males of *G. thoracica* (Smith 1857) and *G. lacteifasciata* (Cameron 1902) compared and contrasted here. The need to identify each species in the genus is crucial to meliponiculture. It is also critical in the accumulation and segregation of its essential products, namely honey, propolis, and pot-pollen. In doing this, the paper will explore the variety of shapes and colours of its nest entrances, nest structure, flight patterns and path, its peculiarities, foraging preferences and affinity. We look at the two species and aim to see how they can be differentiated.

Some experiments were conducted to improve honey pot size by creating wax moulded starter pots in their nests. The pictorial results showed accordingly.

Keywords: *Geniotrigona thoracica*, Honey, Involucrum, Meliponiculture, Nest entrance colours, Propolis.

4.2 Introduction

It is among the largest stingless bees in the region. The meliponiculture of *Geniotrigona thoracica* is very much sought after because of the misconception that a larger bee will produce more honey. Although it is a known fact now that many parameters govern its honey production, one cannot dismiss the idea that bigger bees will make bigger pots.

And therefore, it should produce more honey. We shall investigate the strengths and the weakness it has due to its size.

It does, however, produce much more propolis and has a quicker capacity to do so. We also studied its preferences in acquiring foraging sources. What stimulates its collection, and how it is distributed in the nest?

Figure 9 Geniotrigona thoracica gender and caste

4.3 Materials and Methods.

4.3.1 Material Study and Placement in the Meliponary

Since 2011, we have gathered wild nests from land development projects over the past eight years, be it construction development or agricultural development. We sought permission from local authorities to save bees in land development projects destined for site clearance. The wild nests in felled logs were kept alongside hive boxes.

The newly acquired log nests are placed in transit storage for acclimatization. White powder (a borax mix) at the base is ants and termites repellent (Jalil 2018).

Figure 10 New arrivals in transit area

Figure 11 Placement under some natural shade and close to a body of water. Once placed in the farm proper, the base of the trunks was smeared with used motor oil or grease to prevent crawling insects like borers, ants, termites and even carpenter bees. This will preserve the wood as well.

4.4 Beescape in *G. thoracica* Meliponary

Figure 12 Various foraging sources provided in the Meliponary

The provision of good foraging sources is crucial. After acquiring more than two dozen hives, we felt that we needed to improve the necessary foraging sources.

All sorts and manner of vegetation were provided, from groundcovers (including flowering weeds, low shrubs to tall shrubs, to fruit trees and Palms like coconut and betelnut.

Although the vegetation required full sun, the hives were placed under shady trees and near a water body to ensure the right humidity was maintained. (Jalil 2016) What is causing this behaviour (Figure 13 foraging minerals at a browning patch of turf

) in thoracica? This thoracica colony is seen cutting grass34 below its hive, and another group of them is swarming nearby an H. itama hive but not fighting.

It could be some mineral or salt residue or some overripe fruit rolled over that area! There is no fruit around that area. We only see them foraging or gathering grass residue when they congregate near that area. Their FIFO (fly in - fly out) activity at the nest entrance also had increased traffic.

[34] Editor's note: Trigona naturally prune and remove vegetation near their nest entrance. Perhaps *thoracica* is also trying to keep the vegetation from covering them too.

Figure 13 foraging minerals at a browning patch of turf

4.5 Nest Structure

4.5.1 Nest entrance tubes

Figure 14 shows some *Geniotrigona* nest entrance tubes collected in Borneo. They are in different stages of growth, estimated between 1 to 5 yrs. of age. The entrance tubes are sometimes found on the ground after being broken off accidentally by wild animals.

Figure 14 Nest entrance tubes pf Geniotrigona nests collected in Borneo.

4.5.2 Nest Structure in a feral hive (Figure 15 Nest Structure of a feral hive in a tree hollow.)

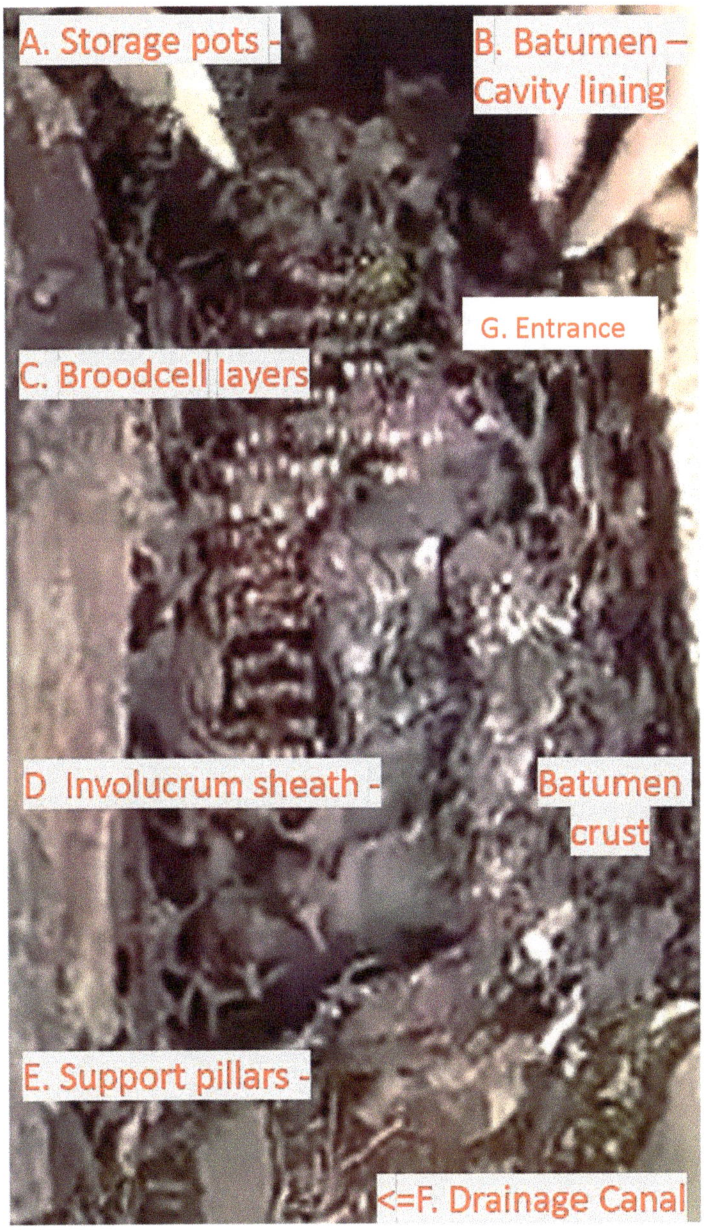

Figure 15 Nest Structure of a feral hive in a tree hollow.

A. Storage pots - for storing honey, pollen and occasionally surplus wax or cerumen.

B. Batumen - a hard geopropolis lining of the cavity wall (the Batumen crust is the exposed batumen flaked or separated from the cavity wall)

C. Broodcell Layers – Egg cells arranged in horizontal layers, occupying the cylindrical cavity of the host tree hollow.

D. Involucrum Sheath – pliable cerumen propolis covering of the brood cells. Usually, to regulate the temperature and humidity of the brood chamber and the queen within.

E. Support Pillars – stiff propolis pillars for support like trusses to keep the brood stable and in place. F.

F. Drainage Canal – a channel for draining accumulated condensed moisture or excess rainwater or other waste fluids.

G. Entrance – The opening that gives access to the external environment. A feature not visible in this image is the entrance tube from outside the tree trunk, which would penetrate the tree's cavity wall. The entrance tube variations are featured in other sections herein.

4.5.3 Nest Structure within a boxed hive[35] (Error! Reference source not found.)

A. Brood Cells - The brood cells are arranged in horizontal layers and sometimes as a spiral comb. It is not just a bunch of cells but organized in a horizontal comb, like in most stingless bees.

Figure 16 A. Horizontal spiralled (anticlockwise) layered brood cells. B. The first thing a colony does is build a thick dark involucrum. C. Initial array of pillars to hold storage pots.

B. Involucrum - The first thing a colony does is build a thick dark involucrum36. This is done when the bees are trying to build an environment for the brood that may not be compatible within the box. As the colony grows, this grows to the side of the walls and ceiling, forming a pliable batumen.

C. Pillars37 to support and hold storage pots. Unlike the cavity in a tree hollow (mostly cylindrical), the top box's cavity is a cuboid. As some storage pots are already attached to the box walls and floor,

[35] Editor's note: The hive is not actually domesticated. The bees in the hive have been placed there (in a box). Their behaviour and ecology, evolved over millions of years, does not then vanish.

Editor's notes:

[36] B Dark Involucrum - The covering of a group of combs? That would make it an involucrum. Otherwise, this is a wall or covering, which would make it a kind of soft batumen.

[37] C. Pillars - The bees have evidently built a few long, skinny pillars. What makes this likely to be the base (or the bases) of storage pots?

the next array of pots will be supported by these pillars. The pillars holding the brood cell layers are thicker.

D. Drainage or Rubbish canal. This does not appear anymore in a box hive as a canal or channel but occasionally appear as small heaps of rubbish or mucus in some locations in the box or any of the boxes of a multi-tiered hive system.

4.6.0 Failures due to weak colonies

Weak colonies may occur when there are insufficient preferential forage sources in the surrounding. The weak colony becomes susceptible to invasive attacks from stronger colonies. Possibility of interspecific competition or even intraspecies competition. (see **Error! Reference source not found.**)

Usually, it occurs due to the inferior colony being are low in population. The low population may also be an accumulation of a few minor faults in the meliponiculture practices itself. Sometimes, the method of transfer from log to hive/box causes the queen to be stressed. This stress then results in the queen's failure in regular oviposition.

It may also be due to several other reasons:

- ✓ Extreme weather conditions causing an ailing colony insufficient resources and time to recover from food and material deficiencies.
- ✓ Stress due to haphazard transfer or insufficient damage recovery period, especially in 'colony division' activities.
- ✓ Box hive walls are too thin and have poor insulation, consequently very stuffy/too high nest temperature or humidity.
- ✓ Incorrect placement of colony or location not suitable. It depends on where we place and maintain the hive for the colony in question to develop comfortably.
- ✓ Pests or predators in the station may also weaken a colony's survival and growth subsequently.

4.6.1 Examples of taken over nests

One particular case of the weakening of a colony of *G. thoracica* is from splitting (a Meliponiculture technique of multiplication). The existence of a young physogastric queen has already been

Figure 17 Examples of weak colonies taken over by a smaller species Lepidotrigona sp.

conforming with a new brood. *Geniotrigona thoracica* workers are not numerous enough to counter an invasion of another stingless bee species (in this case, *Lepidotrigona terminata*). The more significant but weak colony species of *G. thoracica* (**Error! Reference source not found.**) was taken over by a smaller but robust species, *L. terminata*. Fortunately, the queen survived and was transferred to another queenless hive.

The suspected invader's swarm origin is from a full *L. terminata* hive box 10 meters away. Hard lesson well learnt, split weak colony should be moved far away from any stingless bee with such tendency. As shown in figure 17 above, this happened to 3 separate *G. thoracica* colonies – although another two could be some other *Lepidotrigona* sp. The *G. thoracica* nests were eventually taken over completely, and the new *Lepidotrigona* sp. colonies built new entrance tubes over the previous *G. thoracica* entrance tubes.

They (*Lepidotrigona* sp.) are smaller in size compared to *G. thoracica*. Still, sometimes it's a question of numbers when a colony is weak and is declining in workers for some reason or other the invading species can take over. They seem to be more robust and adventurous than the other stingless bee species. They do have a peculiar habit of abandoning their hive to invade other packs (sometimes due to heat or predator threats). This generally occurs when they are out to reproduce or colonise new nests. Other weaker colonies or nests in poor health become targets when there is a lack of new empty cavities.

We need to scrutinize and make available the plants and flowers each species needs, and they will forage harmoniously. They only try to rob others when they cannot get the particular flowering plant that they want[38].

Nectar source is a limiting factor, although any of the hive's nest one buys if the basis is limited, there will be problems later on, and they will attack each other[39]. We advise acquiring some nests first and then gradually add on from time to time and provide enough food for the bee before adding more colonies?

4.7 Morphology - Comparisons between *Geniotrigona* and *Wallacetrigona*

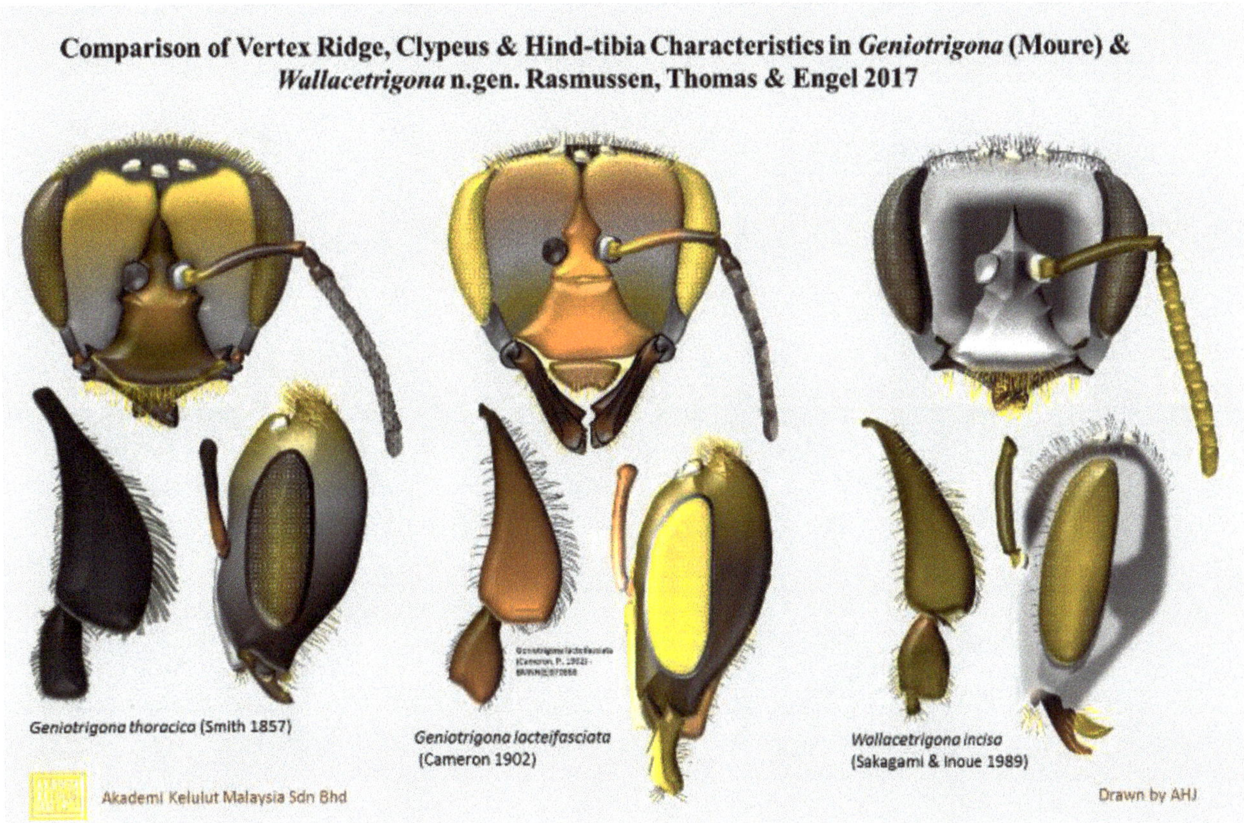

Figure 18 Comparison of Geniotrigona spp. with Wallacetrigona sp. - one of the reasons G. incisa is now placed in a new genus on its own. i.e., Wallacetrigona.

Editor's notes:

[38] The other hypothesis is that pillaging happens when there are odours released from the nest, which signals that it is vulnerable and ready for exploitation. The tendency for aggressive behaviour to be expressed by one species toward another usually is simply termed 'robbing'. There are two professional robbers among the meliponines, mostly in the Neotropics with another in Africa. An interaction between your two species can mean that either might attack the other but beyond that, it really does require some further study.

[39] What is suggested here, is that a colony that is healthy and stable needs to stay that way, and have food and other resources nearby, or there can be problems with pillaging (robbing).

In Rasmussen, Thomas & Engel 2017, a new characteristic was highlighted in the Genus *Geniotrigona*. Comparison of the *Geniotrigona* spp. Shows a raised ridge on the vertex behind the ocelli of *G. thoracica* and *G. lacteifasciata* but not in *G. incisa*/ Henceforth, this is one of the reasons *G. incisa* is now placed in a new genus on its own, i.e., *Wallacetrigona*.

Initially, we thought this characteristic would be a setback in that for *Geniotrigona* spp. The peripheral vision would be obscured[40]. This we presumed earlier may make them particularly vulnerable in battle situations. However, this requires further studies.

4.8 Honey production support trials

Supplementary building material can be provided to support a colony. Some trials were initiated with artificial storage pots to nest structure construction and quicker honey production. Using *Apis* sp. wax to form or mould by shaping from the ends of kitchen utensils handle back end. Materials like Apis wax are bought from other farms and mix with propolis that's in the nest.

The left photo below shows natural storage pots in a nested structure. The middle image below shows an example moulded storage pot made of *G. thoracica* cerumen mixed with honeybee's wax making it pliable and mouldable. The photo below on the right, an example of an arrangement that would be

Figure 19 Left: Natural honey storage pots unruly with 500-rupiah coin for scale indication, Middle: Actual thumb-size artificially moulded pot. Right: Arranged to occupy maximum space and for ease of expected harvesting.

ideal for honey harvesting as compared to the natural randomly placed and overlapping one another.

[40] Editor's note: Their ocelli are not for rear vision; they are for orientation. The compound eyes function to detect motion and patterns. The insect brain integrates and interprets them, and there may be resultant behaviour. Furthermore, bees can move their head and when in flight, would have no problem detecting what is around them.

4.8.0 Results and discussion

4.8.1 Meliponiculture

The hive box's height is quite critical in the sense that if it is too shallow, the box can be filled with propolis and involucrum within one month with little space for storage pots. If too deep, the bees find it too inefficient to regulate the optimal internal temperatures.

An internal height of 10cm to 15cm x B4 paper size (25cm x 35.3cm) seems appropriate, but sometimes this hinders regular harvesting as they build up batumen at the ceiling (be it wood or observation plastic sheet) of the box. Having the boxes stacked vertically up to 3 tiers seems to aid in the colony development and facilitates honey harvest. (Jalil 2016)

4.8.2 Intervention trials with artificial storage pots

Figure 20 A topping box laid with artificial plastic pots.

As shown above in Figure 20, "A topping box laid with artificial plastic pots[41]." Usage of artificially engineered storage pots to assist in honey and pot-pollen production has been experimented with before. Some have sought to use plastic cups as storage pots for the bees. We do not advocate using plastic pots as we fear the low pH of honey may have minute adverse reactions with many polymer-based products. Although there are many types of plastic or polymer-based products, we need to use food-grade plastics as far as possible. However, it is a matter of individual preference! Artificial plastic pots can affect honey quality. Some say the honey will have a plastic taste, and some suggested using glass instead. However, the glass will add too much cost, make it even more challenging to harvest

[41] Editor's note: Anywhere else in beekeeping this would be called: "A honey super equipped with artificial storage pots"

propolis, especially if one has hundreds of colonies! This procedure also hinders the harvesting of propolis.

Example - Polyvinyl Chloride (PVC) is a polymer product that may degrade Honey. PVC should be avoided for food storage. PVC tends to release the chlorine component over time. And it may not be suitable in this instance.

PVC's reported as hazardous when it has direct contact even with water. Since honey (especially Meliponine honey) is slightly acidic, it's not good to expose it, especially with PVC and many other types of plastic. These elements have leached from the plastic into the honey detected through a machine (gas chromatography). Here's scientific literature that discusses the toxic by-products released by PVC. just go through the summary of findings on page 5 in http://mts.sustainableproducts.com/SMaRT/ThorntonRevised.pdf (Thornton, 2000)

The Food Department at Mardi had researched plastic-contaminated honey.

We presumed the bees would overlay with propolis on the inner side of PVC "man-made pots". If not, it may pose a danger to consumers. But you might want to keep your hive box free from toxic gases released by PVC or untested plastic (always opt for food-grade plastics)[42]. And those gases are readily released at room temperature range by the time the bees load up the honey into the cup. And the time consumed by lining the cups takes extra work hence extra time. It becomes redundant because the whole aim is to help the bees reduce time.

Artificial pots are only useful when there are nectar flows, and there is a technique to stimulate the use, similar to the use of artificial propolis/wax pots[43]. The picture below is an early attempt as an alternative to the use of plastic last year. Trials were done with varieties of propolis to wax ratios.

Editor's notes:

[42] A lot of commercial honey is sold in plastic containers, which may not be harmful or inferior. What is important is the kind of plastic used.

[43] I'd say the bees would tear them down and throw them away if they are not natural ENOUGH. This is the beauty of normal cerumen. It can be used to make a variety of things and continues to be workable for months. A non-recyclable container, in a fixed nest space, is a liability.

The other objective is to train the next generation of workers to build slightly larger pots successfully, but after a couple of life cycle, the natural size pots prevails[44].

Figure 21 First trials of wax propolis moulded storage pots randomly placed to observe whether the bees accept them. What was placed was a replicated bottom half of the storage pots D, Pots (circled in red) being built up from the bottom half that was provided earlier

4.9 Conclusion

The end conclusion sees that although it may be quite interesting how we may train bees to accept changes or alter their behaviour[45], it's better to let it go naturally. The accelerating adaptation technique and knowing the most massive pot size threshold acceptable to the colony are feeding the target colony with its honey/pollen. Here is where there is anxiety manipulated. This is the story about artificial[46] pots. If there is no urgent need, it is better to go natural.

Finally, we concur with some advice provided by Bees for Development Journal 110 in March issue 2014:

Reduce your intervention and follow the following three principles.

1. Do not put anything into a nest that does not come naturally from the bees.

---

[44] A bee innately knows what size its storage pots are. There is no question that this is genetic. It differs widely between genera, and also between species of the same genus. If hybrids were possible, it would probably be shown to show intermediate sizes.

[45] Editor's note: If you had a preliminary observation, then it might be worth pursuing.

[46] The authors are also advocating the avoidance of plastic materials due to the long-term adverse impact on the environment and the non-biodegradability of plastics.

2. Do not take anything from a hive that they cannot afford to lose.

3. Allow yourself to be guided by the bees. Love your bees (my add-on)

To be guided by the meliponines, you need to familiarize yourself with their behaviours and needs: this is something you can learn from the observation and by interacting with experienced stingless bee breeders.

(excerpt from BfD Journal 110 Editorial note March 2014 by Nicola Bradbear)·

Review of these man-moulded cerumen storage pot test results to increase production capacity, video session 1 (beginning) click ☞ https://youtu.be/kcorosxd_kc video session 2 (1-month outcome) click ☞ https://youtu.be/sacxkmzp4mq (still images Figure 22 Progress of moulded pot trials for a medium-

Figure 22 Progress of moulded pot trials for a medium-sized species.

sized species.

below)

Chapter 5

5.0 Concoctions of Meliponicultured Honey and Traditional Herbal Medicines Developed from Correlations with Ethnoentomology in Langkawi, Malaysia

Map 1 Map of Langkawi Island

5.1 Meliponiculture in Langkawi, Malaysia

The year was 2014 when we first visited the bee farm of Mr. Radzi Alias in Northeast Langkawi's suburbs.

Figure 23 A. from left: Dr. Ibrahim Shuib, AHJ, Radzi Alias and Yusnizar Ibrahim (Photographer); B. & C. Heterotrigona itama entrance activities, D. Discoloured propolis possibly due to fungus; E. & F. Hives under shade nets; G. Hives destroyed by Megachilid bees after massive cerumen pillage.

He started beekeeping at the farm when he implemented an integrated farming system called TRAC Langkawi, i.e., Training Retreat and Agriculture Centre Langkawi, in 7 acres.

In October 2012, he began to take meliponiculture classes in Kelantan, then by December that year. He joined a program with Kelantan Biotech in Min Camp at Kubang Kerian. After that, he started a full-time career with kelulut.

Figure 24 A. & B. Healthy H. itama brood; C. Tetrigona apicalis entrance hole at box hive; D. Tetragonula laeviceps box hive

He started at the end of the year in Langkawi after picking up stingless beekeeping techniques in Kubang Kerian, Kelantan. He has been teaching village folk to locate colonies in the mangroves, and in three weeks, they could supply him with 60 colonies. At that time, he was learning to transfer colonies into boxes and then by 2013, he had made colony transfers to more than 200 hives. The first six months' development is profitable because he could harvest honey 8-10kg of honey every month. That time he had a honey stock of 100-200kg by the end of 2013. There was a bit less production of honey in the boxes.

Figure 25 A. Meliponary sign boards sponsored by Govt. Agency; B. Understorey of the coconut farm; C - G. Various sections of the farm with creative roofing. Yellow painted zinc roof & aluminium lined polycarbonate roofs with new dimension (19" x 32"). Overly large size dimensions to ensure sufficient coverage from rain and sun; H. Mr. Radzi in the new Meliponary.

Figure 26 A. & D. Harvesting equipment (Honey suction pumps); B. Full Harvesting gear; C. H. itama honey pots; E. & F. pollen pots; G. Tetragonula laeviceps storage pots and cells; H. & J. Extraordinary Large honey pots of H. itama.

Figure 27 The Meliponary in a coconut farm (pandan variety)

In March 2014, he started to place a box (topping) above the log. Then during 2015, a box-like "Mustafa hive" was attached to most colonies. However, by then, he experienced significant cerumen pillage attacks from Megachilids. He lost about 70-80 logs and hives in boxes. By the end of 2015, there were only 170 colonies left. He then decided to move to a new location in a "Kelapa Pandan" farm in Durian Perangin.

Figure 28 Brown Sea cucumber (or "deep-water redfish"), *Actinopyga echinites*. used in the concoctions.

He observed some of the villagers' traditional practices that use Honeybee (*Apis*) honey by setting up there. Combined traditional herbs with *Apis* honey produces some popular concoctions. Using *gamat*, local herbs such as 'Pepper Elder' (*Peperomia pellucida*) and garlic and ginger, lemon and a mixture of honey with 'Tongkat Ali' (*Eurycoma longifolia* Jack) are a series of products from Black Bee Farm. Some of the traditional practices that use *Apis* honey by the villagers for the local residents was replicated with stingless bee honey. For example, use sea cucumber, local herbs and garlic and ginger. *Eurycoma longifolia* is a product range from Black Bee Farm.

Figure 29 Peperomia pellucida, Pepper Elder, locally known as 'sinaw-sinaw' in the Philippines (literally 'shining')

a. The Sea Cucumber (Gamat (*Actinopyga sp.*)) is washed, and the innards are removed, then cleaned and soaked into honey. After two weeks, chopped, then put back into honey and be used for medical purposes or supplements. The most popular applications involve spinal disorders, back problems, urinary tract, and the body that is quite sluggish. The effects of the body are very noticeable after taking it, especially the system of swelling, which means increasing the metabolism rate of the body.

Figure 30 Sea cucumber maceration in Honey

b. Garlic, ginger and lemon honey. These three ingredients are cooked and then cooled. Newly blended with honey and stored. This supplement helps in cholesterol and congestive heart problems.

c. Cucumber honey (*Peperomia pellucida*) & halba (Fenugreek) - whole plant & shoots washed and cleaned then boiled with fenugreek, cooled and mixed with honey. Many helpful for cough problems, sore joints and gout problems.

d. Tongkat Ali's sweet honey (finely ground *Eurycoma longifolia* and macerated in honey) is in demand for the goal of manhood (erectile dysfunction aliments).

e., Garlic honey. (Garlic is burned and soaked in honey for two weeks.) It is suitable for prolonged fever problems and a body with less disease-resistant data.

Take on normal condition as supplement 1-2 tbsp daily empty stomach time. For those who, for treatment purposes, recommended 2-3 times a day.

For added libido, we recommended consuming in the evening. Honey gamat (Sea cucumber cleaned, and the stomach removed, and the meat is chopped and then cooked and mixed with honey)

The Black Bee Farm Langkawi management now supplies Hotel St. Regis (https://www.facebook.com/stregislangkawi) 5 - 10 kg a month from Oct. 2016. The hotel provides Madu Kelulut as a dessert (dessert)

5.1.1 Testimonial from Hotel St. Regis, Langkawi:

Quoted by Executive Chef Gaetan Biesuz Hotel St. Regis, Langkawi

"Do you know the dessert of Black Bee Honey in Kayu Puti's new menu is produced in Langkawi?

Locally known as 'Lebah Kelulut' or Stingless bees, the honey tastes are sweet, sour and sometimes bitter yet pleasant taste to tantalise your taste buds. The honey contains health benefits such as strong anti-bacteria, anti-toxin and natural antibiotic elements.

When I first tasted the local honey here, I was caught by surprise because honey always meant sweetness, but this is unique."

Few other 5-star hotels have been approaching the Black Bee farm to place some hives in their garden to attract their guests. However, the maintenance programs for the Hotel Meliponary were finalised.

In addition to these programs, the management of Black Bee Farm is also in cooperation with some cosmetics development parties, and efforts are being made to introduce Stingless Bee products to other countries

Figure 31 A. Harvested Pollen (pot-pollen); B. Selected herbs for cosmeceuticals; C. Packaged products; D. Assorted cosmetic products.

5.2 Taman Kelulut Kompleks Kraf Langkawi

Figure 32 A. Meliponary owned by Mr. Nizam at the Handicraft Centre in Teluk Ewa, Langkawi; B. Mr. Nizam with his hives under the Raintree; C. & D. Ornate carved stingless beehive boxes; E. Experimental design of local roof for hive box; Tetrigona apicalis nest entrance.

Mr. Mohd Nahizam b Nosrudin set up a farm in 2015. It was initially a hobby but lured by the potential income, and he worked hard on his home ground. However, a year in farming activities, he faced black soldier flies (BSF- *Hermetia illucens*)[47] threat. He suffered destruction on more than half the stock of his colony. In this disaster, there is fate in which he has had the opportunity to start a meliponiculture setup at the Kompleks Kraf Langkawi, Teluk Yu, Mukim Bohor,07000, Pulau Langkawi, Kedah Darulaman, Malaysia.

Here are hives placed under the shade of raintrees (*Samanea saman*). At an adjacent spot, he constructed a structure of black netting as added shade. He is working with the craft's centre in fabricating boxes of the nest box with ornate carvings.

The honey harvested is now sold at the counter in the handicraft centre.

Figure 33 A. & B. Hives under raintree; C. & D. Hives under netted shade structure

Product: The honey harvest is now sold at the counter in the Craft Complex.
Large bottle 200gram @ RM 68.00; Small bottle 115gram @ RM 40.00 and propolis soap sell for RM 20.00 each
The Langkawi Department of Agriculture financed the labelling.

[47] Editor's Note: They are presumably the same we have in the Neotropics, which lay eggs that produce large, armoured larvae that decimate the brood and storage pots..

Figure 34 The main products and by-products from Madu Kelulut Resources.

5.3 Some traditional practices that use honey

- **The Sea Cucumber** (Gamat (*Actinopyga* sp.) The effects on the body are very noticeable after taking it, especially the system of swelling, which means increasing the metabolism rate of the body.
- **Garlic with ginger and lemon honey**. This supplement helps in cholesterol and congestive heart problems.
- **Cucumber honey** (*Peperomia pellucida*) & halba (Fenugreek) are mostly helpful for cough problems, sore joints and gout problems.
- **Tongkat Ali's sweet honey** is in demand for the goal of manhood (erectile dysfunction aliments).
- **Garlic honey**. It is suitable for prolonged fever problems and a body with less disease resistance.

5.4 Traditional & Indigenous knowledge Vs Scientific Analysis & Findings

- To better inform the public and scientists on the validity of such claims,
- Traditional or indigenous knowledge needs scientific data as an endorsement of their practices.
- Therefore, this collaborative research aims to test and evaluate these traditional practices that incorporate local herbs for traditional medicinal purposes.
- This work's novelty lies in the potential findings of beneficial metabolites or compounds in the honey and herbs mixtures.

5.5 The MGI team for Data Analysis of the Honey concoctions from Langkawi Malaysia

Dr. Nor Azfa Johari - GCMS expert

Mohd Razif Mamat - Bioassay expert

Dr. Azira Muhamad - NMR expert

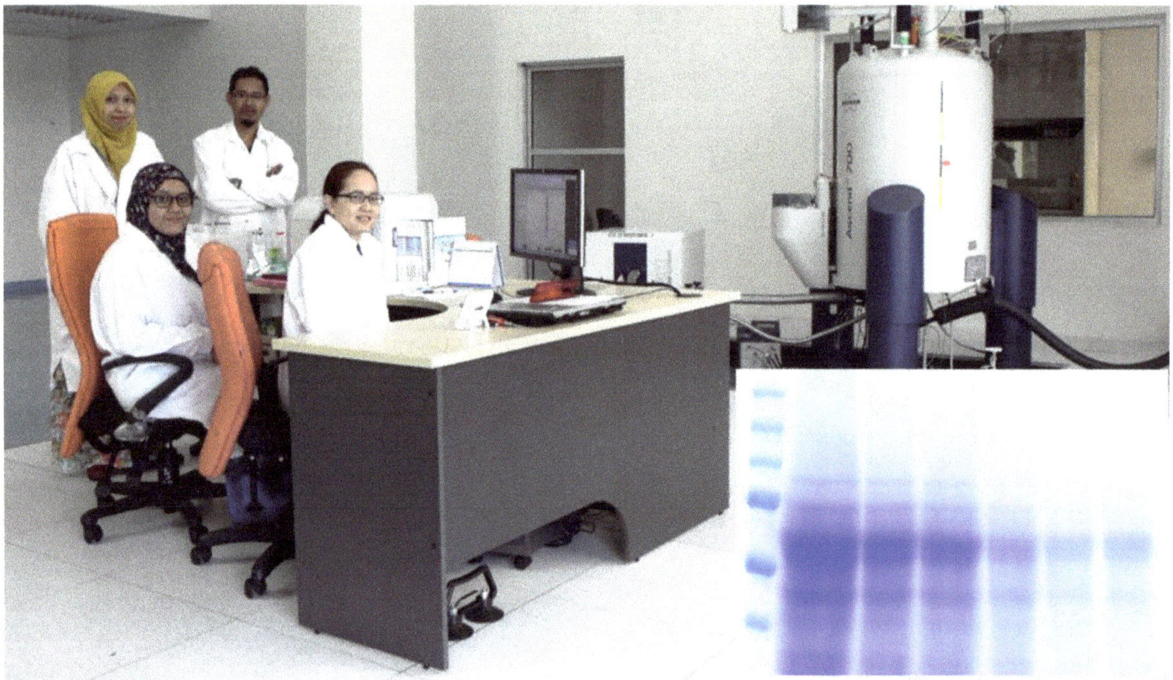

Figure 35 Metabolite profile analysis using Nuclear Magnetic Resonance (Bruker Ascend 700 MHz); inset: Electrophoretic Pattern, Chromatogram profile

5.6 Research Aim

1. In this research, we propose to profile the metabolites in each of the preparations or mixtures of honey with four different local herbs.

2. This profile will include the sugar, amino acid and organic acid contents identified through the metabolomics approach using Nuclear Magnetic Resonance (NMR) spectrometer and Gas Chromatography-Mass Spectrometer (GCMS).

3. We will analyze and distinguish the active compounds that are beneficial in the macerated mixture from the data.
4. We will look into the antioxidant properties of the compounds identified.
5. Cytotoxicity study using mammalian cell culture will also be carried out.
6. The produced data are expected to support or otherwise, the indigenous knowledge in traditional treatment by providing test results to these practices[48].

Objectives

1. To collect and process the herbs and honey mixtures
2. To determine[49] the beneficial compounds by performing metabolomic NMR and GCMS analyses
3. To analyze the function of the herb mixtures by conducting a bioactivity assay
4. To provide data for local practitioners using herbs as a medicinal treatment.

Justification

Four different types of local herbs and honey mixtures were chosen and typically used as a traditional treatment by the local villagers on the island of Langkawi. This research will employ metabolomics as they are excellent methods to identify compounds and metabolites of the honey and herbs mixtures. Antioxidant properties and cytotoxicity studies of the combinations will help test the function of the remedies and correlate to the human body. These preliminary data can form a basis to conduct a further clinical trial.

Expected outputs

Primary and secondary metabolites were identified using NMR spectrometer and GCMS methods. Furthermore, antioxidant properties are expected to be present in these herbs' mixtures. These findings will provide chemical tests and cultural and medicinal testing protocols to use local herbs for traditional medicinal purposes. Practitioners will be able to use the findings as a material data sheet for their herb mixtures.

Editor's notes:

[48] They will attempt to test for the activities of some components used in the traditional practice. There are other aspects of all medicines, and these are cultural, i.e. what the witch doctor and you, and your group, choose to believe. Science likes to see components but, as ever, in the human milieu it is not that simple...

[49] one identifies the compounds, one then seeks to test them, with a suitable group and protocol.

5.7 Activities & sampling

1. Sample collection & processing in Langkawi

i) sea-cucumbers (gamat) mixed with stingless bee honey

ii) 'Pepper Elder' (*Peperomia pellucida*) mixed with stingless bee honey

iii) garlic, ginger, lemon and stingless bee honey mixture

iv) honey mixed with 'Tongkat Ali' (*Eurycoma longifolia*)

2. Profiling of metabolites by NMR at Malaysia Genome Institute

100 mg of the mixtures dissolved in methanol-d4 and deuterated water to the final volume of 600 uL. 1D 1H NMR experiment was carried out. Spectra will be analysed using Chenomx software.

3. Profiling of volatile compounds by GCMS at Malaysia Genome Institute

1 g of the mixtures will be extracted using methanol and fractionated by dichloromethane and water before the extracted medicines are injected into the GCMS.

4. Activity assay at the Malaysia Genome Institute

Antioxidant assay and cytotoxicity study using mammalian cell culture will be carried out.

Time frame	Oct	Nov	Dec	Jan	Feb	Mar	Apr	May	Jun
Sample Collection & Processing									
NMR Profiling									
GCMS Profiling									
Activity Assay									
Report write-up									

5.8 Potential Contribution to The Research

Good Health and Well-being: This research will provide scientific proof in product manufacturing based on traditional practices. Indirectly consumers will understand the benefits of the honey mixture when practised efficiently. Thus, the well-being of consumers will improve.

Decent Work and Economic Growth: The scientific data produced will give consumers confidence. Indirectly users will buy the item at a constant rate. In the event of a transaction, the seller's revenue will increase. Hence will indirectly improve the social economy of meliponiculture.

Industry, Innovation and Infrastructure: Supports practitioners to innovate using a mix of different herbs with evidence provided by scientific data.

Chapter 6

6.0 Unique Stingless bees of Penang Island, Malaysia

By Abu Hassan Jalil

This chapter holds sentiments from my childhood. The year was 1967, and I had finished primary education. My family was living in Georgetown, Penang Island. We resided in a government quarters barracks-style double-storey house in the street called Jahudi Road. Jahudi means "Jewish" or "of Jews" in Malay. Not that it was a street for Jews, but there was a Jewish cemetery somewhere in the central portion of that road, and the home I stayed in was about one block of houses away from that cemetery. Although the no trespassing rule was strict, the neighbourhood kids, including myself, frequented the cemetery for our "war games" with makeshift wooden rifles.

Figure 36 The Penang Jewish Cemetery

The cemetery was bordered by a 2m tall by half a meter-thick dark dried-moss covered brick wall with cracks and broken glass pieces cemented on the top in place of barb wire. This was guarded by an elderly "Rabbi" whom we called Pak Rabu. He had an Indian lady companion, and they lived in a small shack in the cemetery compound. Their son, about my age, was the one who frequently invited us for these "war games" without Pak Rabu's knowledge. This tale's significance was the surrounding wall and the tombstones, which were our imaginary war bunkers in our games. Now in the cracks of the wall and some tombstones, there were small dark insects which we only assumed were "Penyengat" or stinging insects. We knew nothing of bees and even less of Stingless Bees.

Jahudi Road, also known as Jalan Yahudi, has since been renamed Jalan Zainal Abidin (https://www.thestar.com.my/news/community/2013/07/13/jalan-yahudi-to-jalan-zainal-abidin-the-real-zainal)

Figure 37 Entrance gate to the Jewish Cemetery

https://pht.org.my/the-penang-jewish-cemetery-site-visit/

Situated alongside Jalan Zainal Abidin (the new name for Jahudi Road), in George Town's heart, the Penang Jewish Cemetery is believed to be the oldest Jewish cemetery in S.E. Asia. The oldest Jewish tombstone is dated 9 July 1835, dedicated to a Mrs. Shoshan Levi and believed to mark the grave of an English woman that was the benefactor who donated the land where the current cemetery stands.

https://www.penangjewishcemetery.com/

Figure 38 Perimeter wall of the Jewish Cemetery

https://www.penangjewishcemetery.com/blog/boundary-wall-restoration

At the same time at my home, on the 2nd storey, there was a veranda. We had “Bertam” (wild Bornean sago - Genus: *Eugeissona*) palm slats blinds at this veranda or, as we called it “, chicks”. The stitched horizontal slats were lined with a navy-blue cotton cloth. It is rolled up or down manually by rope and small pulleys to protect the veranda from lousy weather or direct hot sun. We had large “Ketapang” (*Terminalia catappa)* trees in front of the veranda, so the chicks were seldom used and were rolled up most of the time during the half-yearly dry season. During this period, I encountered these dark insects that seemed to nest in these rolled up chicks. We only knew them as some bees, so we were very wary of them.

I was enrolled for my secondary schooling in Penang Free School. (https://en.wikipedia.org/wiki/Penang_Free_School)

https://commons.wikimedia.org/wiki/File:Cmglee_Penang_Free_School_field.jpg

https://upload.wikimedia.org/wikipedia/commons/f/f9/Pg_botanic_gardens_trees.JPG

Figure 39 Penang Free School

Figure 40 The Penang Free School field

The Penang Free school had a large playing field where we had our P.E. lessons and football or soccer and rugby practices. We also had cadet marching band practices. The field was bordered by large Angsana (*Pterocarpus indicus*) trees around the perimeter. I remember the older boys toying with some insect's nest at the trees. The insects had protruding flexible tubes, and the boys would clamp them with twigs like chopsticks and be fascinated that it would return to normal the next day. We did not learn that they were stingless bees then and just wrote it off as some peculiar insect phenomenon.

Figure 41 The Penang Botanical Gardens

https://en.wikipedia.org/wiki/Penang_Botanic_Gardens

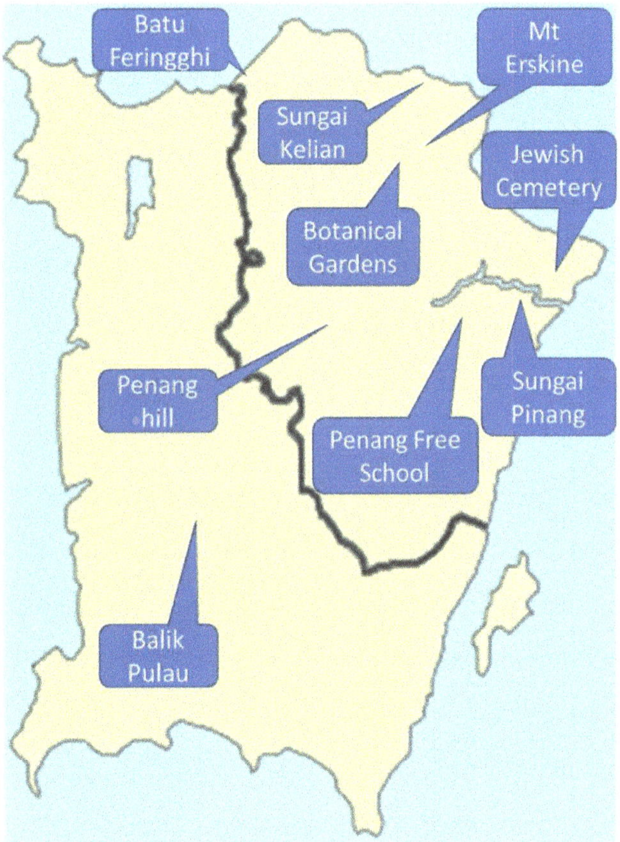

Map 2 Significant locations on Penang Island, Peninsula Malaysia

As a kid, I was incredibly excited to go on excursions fortnightly. That's when my father would bring the family to the Botanical gardens or the beach on alternate Sundays.

The Botanical gardens was just a place to romp and play with some aerial toys. We knew about the "Birds and the Bees" but knew nothing of Stingless Bees then. We did see some bees' "nests" in the large old trees but always veered away for fear of bee stings. Reminiscing those memories with the current knowledge of bees, it is sad for the opportunities missed.

On alternate weekends we would go for picnics at the beach, and my favourite spot was Batu Feringghi. (In Malay, "Batu" means rock and "Feringghi" mean Portuguese) I remember tiny dark bees hovering around the sandwiches that we brought.

I am just recollecting the locations I visited as a student in Penang. There was the Penang Hill recreation garden which is a tourist spot now. I also visited durian farms in Balik Pulau, and both these places had tiny dark bees quite common then, in the 1960s. There was no clue then that these were the stingless bees that would be such a popular trend in these times after more than 50 years.

Out of familiarity with the Island and subsequent curiosity, I wondered why the Island had such a historical significance in discovering those stingless bees. So much so that one species was named for the island. There are at least a dozen mentions of this locality in published historical annals and records of Indo Malayan stingless bees. This implies that those Pre-WW1 European naturalists, zoologists and entomologists favoured flocking on to this tiny "Pearl of the Orient", as Stamford Raffles once termed it.

Now we know that at least Theodore Dru Allison Cockerell (1866–1948 and incidentally a Jewish name) would be in Penang a century ago. Theodore Dru Alison Cockerell was an American zoologist born in Norwood, England. Penang was then a British colony, and it is no wonder that it would be a destination in his numerous expeditions. I would love to imagine that he was probably visiting a Jewish relative in Penang at that time.

6.1 Tetragonula penangensis **(Cockerell, 1919)**

Trigona penangensis **Cockerell** 1919c: 78, 79: **Holotype** (AMNH) (comparative notes, the key to species); **Type locality: MALAYSIA** "Penang Island (Baker 9075)" (worker); *Tetragonilla penangensis* (Cockerell 1919) Tegula black (Penang)....-. *penangensis* sp. nov. (9075)
The wings of *T. penangensis* are dusky, although less produced apically than those of *T. iridipennis*. And not as dark as *T. biroi*. The abdomen is pure black and body length, not over 4mm. (Cockerell, 1919).

penangensis, Tetragonula (Cockerell, 1919) [Typ. loc. : Malaysia "Penang Island (Baker 9075)"]# (Rasmussen, 2008) = Refers to Penang + ēnsis (see suffix –ensis). A State in western Malaysia which has Georgetown as its capital., Island in Penang state. Founded by Francis Light and named after the Pinang Palm (Betel Nut Palm - *Areca catechu*).

Tetragonula penangensis (Cockerell, 1919)

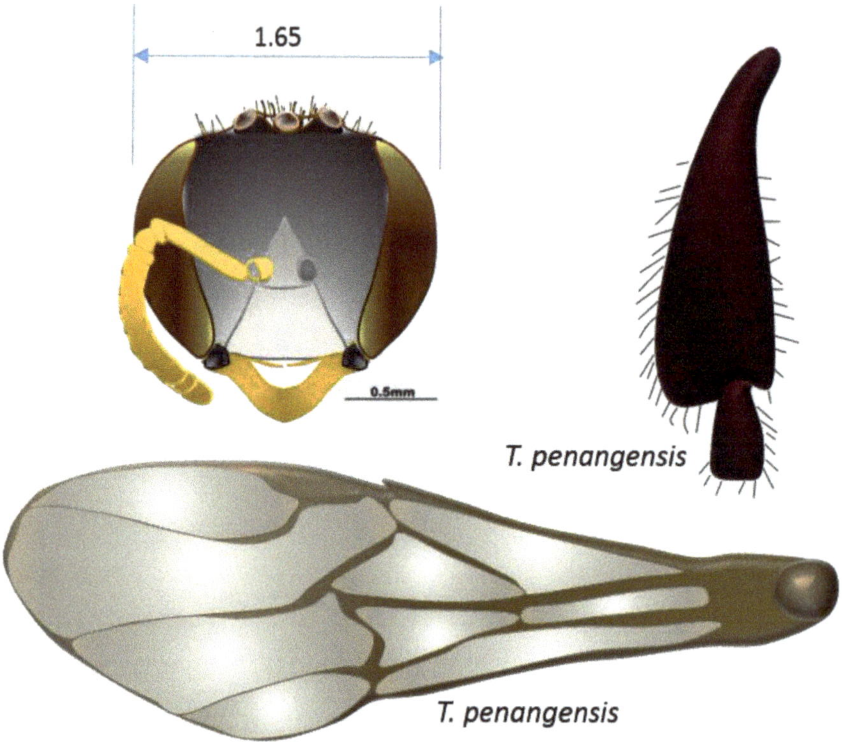

Images Ref:
https://www.discoverlife.org/mp/20q?search=Tetragonula+penangensis&flags=col2:subgenus:

Illustration 1 Tetragonula penangensis (Cockerell, 1919)

6.2 *Tetragonilla atripes* (Smith, 1857)

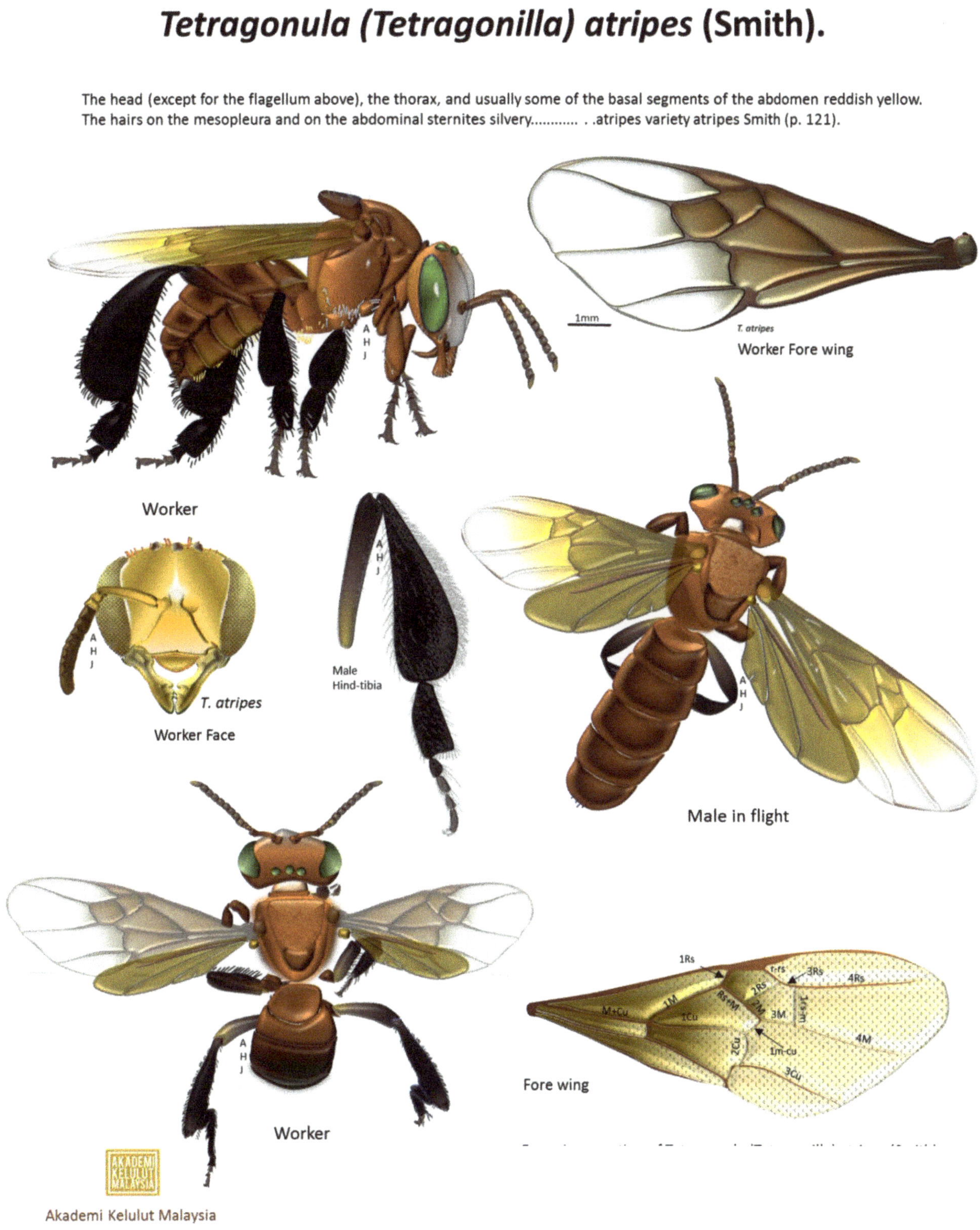

Illustration 2 Tetragonilla atripes (Smith, 1857)

Trigona atripes **Smith** 1857: 50: **Type** (OUMNH (=Wilson Saunders collection)) (taxonomy); **Type locality: MALAYSIA** "Malacca (Mt. Ophir)" (worker); Yoshikawa et al. 1969: 173 (distribution); MALAYSIA, George Town, Penang; Cockerell 1919 *Trigona atripes* Smith, a variety differing a little in the colour of the legs, comes from Penang Island (Baker 9068). Baker on Penang obtained the following species with a black head and thorax.

Smith described *T. atripes* found in Penang. It is fulvous, quite unlike others.

(a) The head (except for the flagellum above), the thorax, and usually some of the abdomen's basal segments are reddish-yellow. The hairs on the mesopleura and the abdominal sternites silvery ....***Tetragonilla atripes*** (Smith 1857) (See Schwarz 1939, p. 121). BORNEO; BURMA-TENASSERIM; THAILAND; MALAYSIA- PENINSULA; INDONESIA-SUMATRA

atripes, *Tetragonilla* (Smith, 1857)[#] (Rasmussen, 2008) = From Latin *ater* ("black") + *pēs* ("foot"). **atripēs** m, f, n, (*genitive* **atripedis**); *third declension* - black-footed.

Friese interpreted as the male of *atripes* (1918, p. 520) and "a variety differing a little in the colour of its legs" reported from Penang Island (Cockerell, 1919, p. 78),

6.3 ***Tetragonula fuscobalteata* (Cameron, 1908)**

Trigona atomella **Cockerell** 1919b: 243-244: **Holotype** (USNM 29467, worker): examined, "Island of / Penang/ Baker", "Type No. / 29467 / U.S.N.M.", "Trigona / atomella / Coll TYPE"); 2 paratypes (USNM) (comparative notes, taxonomy); **Type locality: MALAYSIA** "Island of Penang (Baker, 9585)" (worker);

Trigona atomella, which is possibly considered a variety of fusco-balteata than an independent species, was described from the Island of Penang and subsequently recorded from various localities in Siam. (Schwarz 1937)

fuscobalteata, *Tetragonula* (Cameron, 1908**).** = fusco (Latin – darken) + balteata (Latin **balteatus** - furnished with a girdle or belt)

Trigona atomella, which is possibly rather be considered a variety of *fusco-balteata* than an independent species, was described from the Island of Penang and subsequently recorded from various localities in Siam (Schwarz 1937)

Tetragonula fuscobalteata (Cameron, 1908)

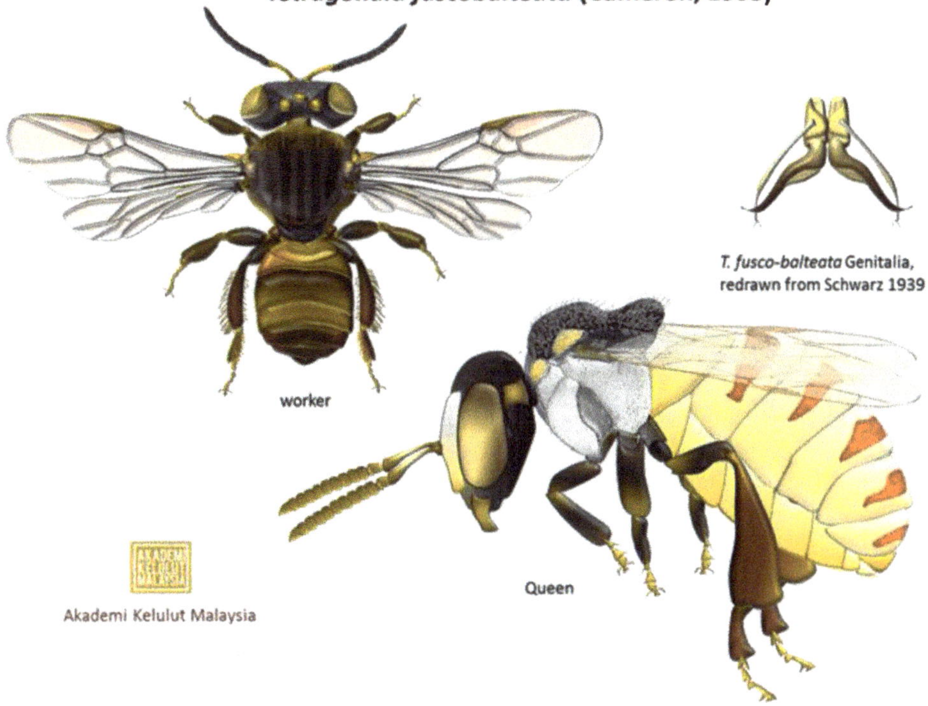

Tetragonula fuscobalteata (Cameron, 1908)

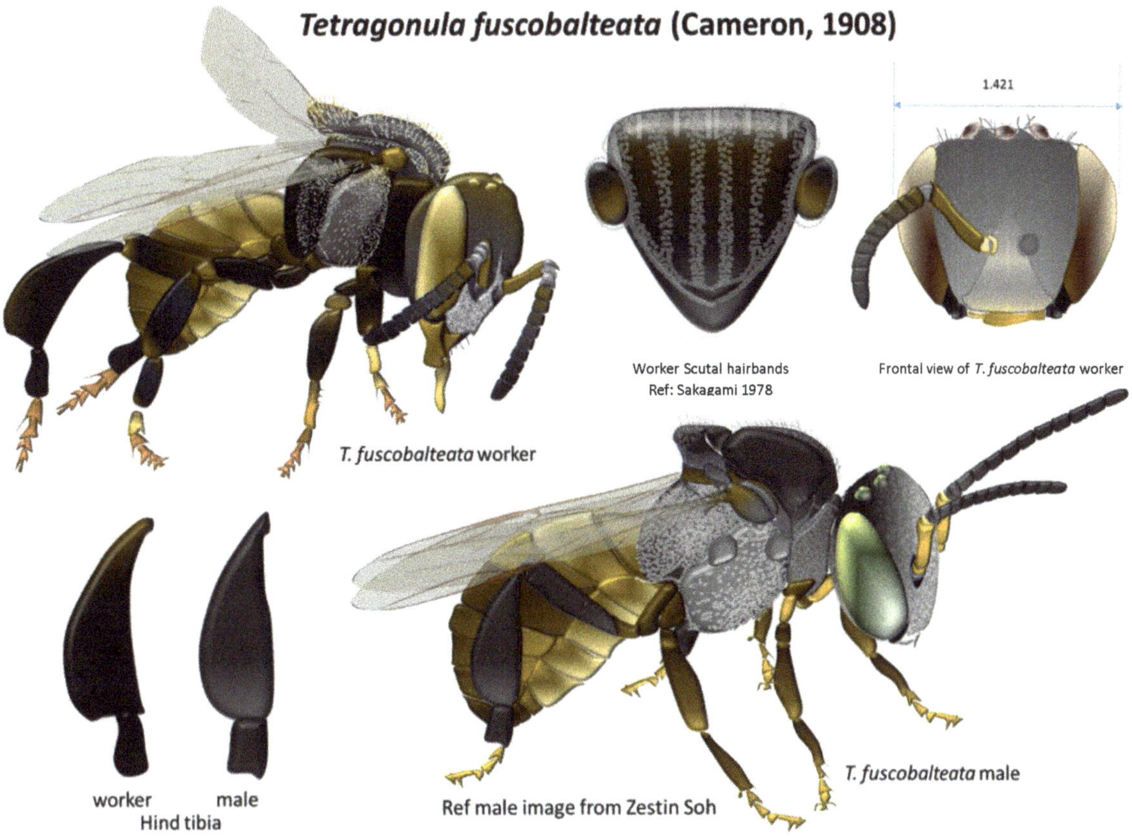

Minute bees, body length 2.8-3.2 mm, length of wing + tegula 3.1 - 3.4 mm; hairs fringing the anterior contour of the hind tibiae usually silvery-gray; color of abdomen very variable, sometimes pallid testaceous with more or less fuscous bands on the tergites, in other cases darker brown .... *Tetragonula fuscobalteata* (Cameron, 1908)

Illustration 3 Tetragonula fuscobalteata (Cameron, 1908)

6.4 ***Tetragonula laeviceps* (Smith, 1857)**

Camargo 1988: 372 (distribution, synonymy); MALAYSIA, Penang;

***Tetragonula* aff. *laeviceps* (Smith 1857)**

7

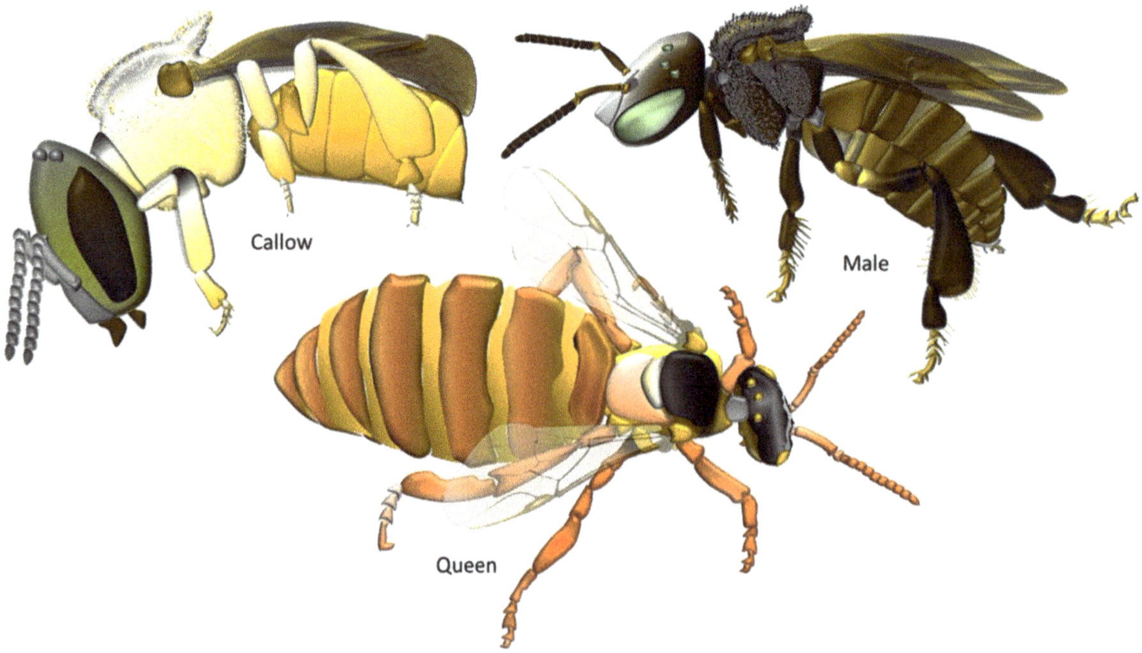

T. laeviceps (Smith 1857)* : It may seem surprising that the identity of *T. laeviceps*, one of the best known names among the Asian stingless bees, is a matter of discussion. Populations called *T. laeviceps* have an extremely large range and there is substantial variation among populations; thus *T. laeviceps* is probably composed of two or more biological species. To complicate matters, the type specemen of *T. laeviceps*, originally collected from Singapore, is lost. Rasmussen and Michener (2010) designated a neotype also collected from Singapore.

Illustration 4 Tetragonula laeviceps (Smith, 1857)

6.5 *Tetragonula pagdeniformis* (Sakagami, 1978)

Trigona (Tetragonula) pagdeniformis **Sakagami** 1978:"Penang" (8 worker); paratypes (BMNH (5)

pagdeniformis, Tetragonula (Sakagami, 1978) - relating to the form of *T. pagdeni*. Ref. *T. pagdeni*- (Named after the naturalists H T Pagden (speciality Hymenoptera) and also in relation with *Trigona* (*Tetragona*) *fusco-balteata* variety *pagdeni* Schwarz 1939 (Rasmussen, 2008))

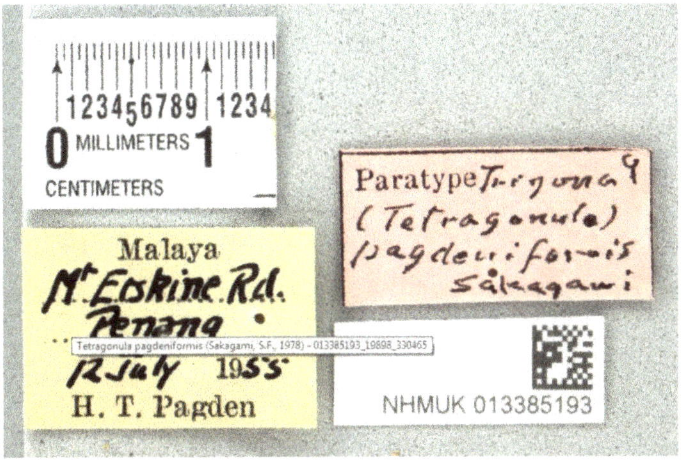

Label 1 Tetragonula pagdeniformis (Sakagami, 1978)

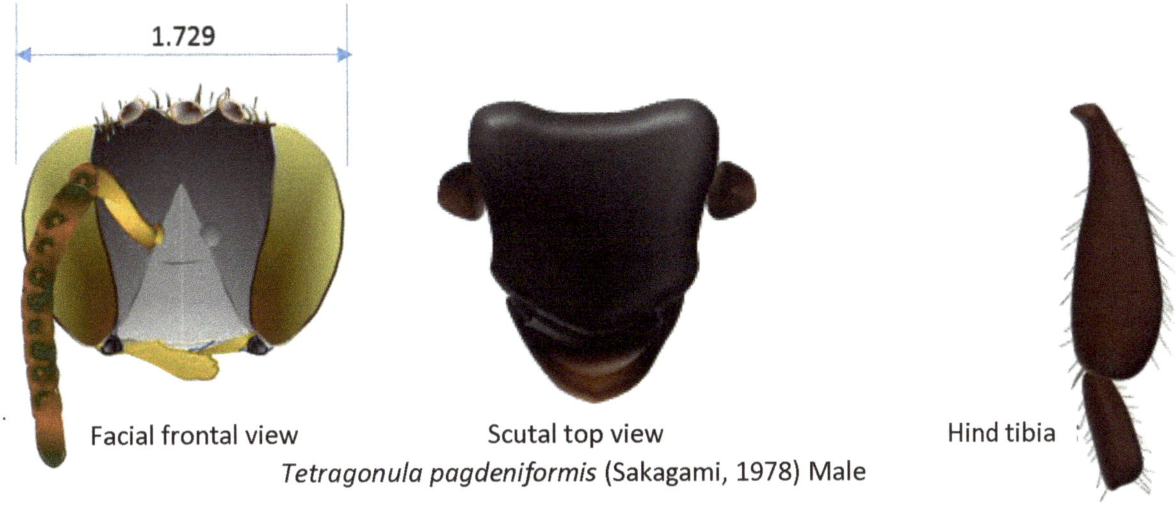

Illustration 5 Male Tetragonula pagdeniformis (Sakagami, 1978)

6.6 *Tetragonula zucchii* (Sakagami, 1978)

Trigona (Tetragonula) zucchii **Sakagami** 1978: 166-194, 196, 197, 198, 208-210*, 228, 229, 236, 237, 238, 247, plate 5: **Holotype** (SEHU) (nest, taxonomy); **Type locality: MALAYSIA** "Fraser's Hill-b, Malaya" (9 workers, holotype, 8 paratypes);"Penang" (1 worker);

zucchii, *Tetragonula* (Sakagami, 1978) – Named after the naturalists Prof. Dr. Ronaldo **Zucchi** and also relates to [*Trigona (Tetragonula)* ***zucchii*** Sakagami 1978][#] (Rasmussen, 2008)

Prof. Dr. Ronaldo Zucchi was a Director for the University of Sao Paulo, Brazil (1984-1988)[50]

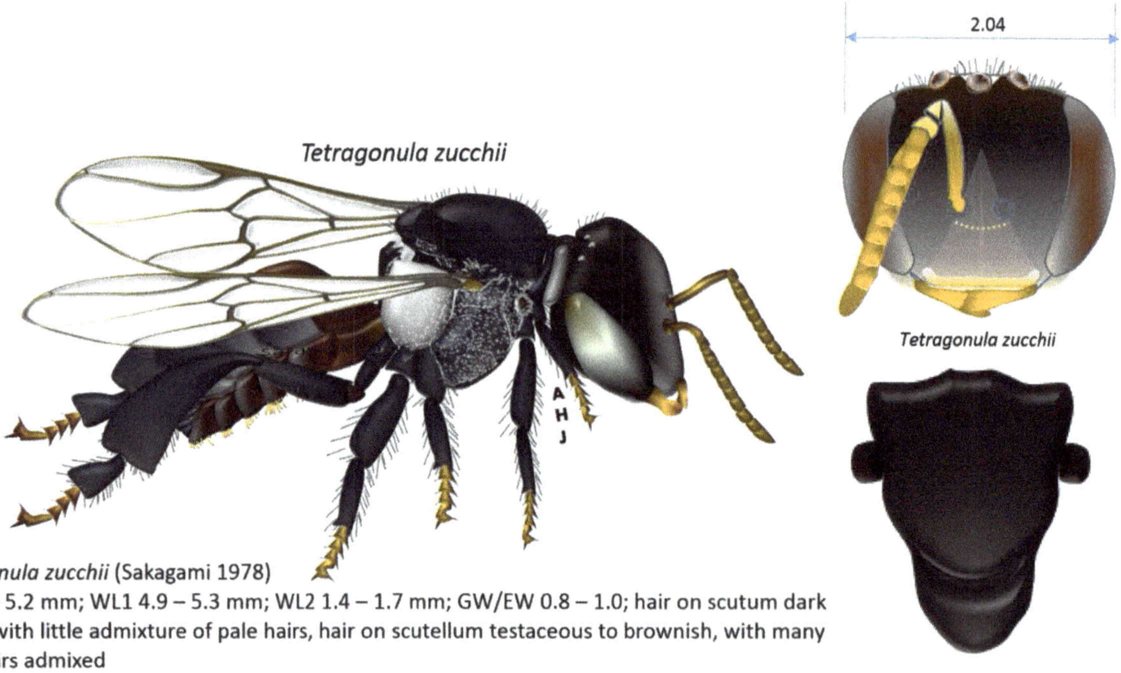

Tetragonula zucchii (Sakagami 1978)
BL 4.4 – 5.2 mm; WL1 4.9 – 5.3 mm; WL2 1.4 – 1.7 mm; GW/EW 0.8 – 1.0; hair on scutum dark brown with little admixture of pale hairs, hair on scutellum testaceous to brownish, with many dark hairs admixed

Illustration 6 Tetragonula zucchii (Sakagami, 1978)

[50] Editor's note: I knew him at USP in Ribeirao Preto, in 1986, and he was not a 'field man'

6.7 *Tetrigona apicalis* (Smith, 1857)

Trigona apicalis **Smith** 1857: 51: **Holotype** (OUMNH); additional putative type (BMNH 17b.1188) (taxonomy); **Type locality: MALAYSIA** "Borneo (Sarawak)" (worker); Yoshikawa et al. 1969: 168, 173, 174 (distribution, nest); MALAYSIA, George Town, Penang;

Figure 42 Tetrigona apicalis (Smith, 1857)

H. bakeri is about 6mm, with the front mesothorax and abdomen shining, and the mesothorax is exceptionally smooth and polished while in *H. itama* distinctly dullish. The legs are black in both (Cockerell, 1919)

Camargo 1988: 370 (distribution, taxonomy); MALAYSIA, Penang;

6.8 *Tetragonula iridipennis* (Smith, 1854)

Trigona iridipennis **Smith** 1854: 413-414: **Holotype** (BMNH 17b.1114, worker): examined, "Type" (orange border), "iridipennis / Type Sm.", "B.M. TYPE / HYM. / 17B.1114", "Trigona / iridipennis / TYPE. Smith.", "Ceylon" (reverse side "53 / 23"). Citations for *iridipennis* are included in a broad sense, as species limits for this taxon is uncertain.

iridipennis, *Tetragonula* (Smith, 1854)[#] (Rasmussen, 2008) = iridi (Latin iridare – "To make iridescent") + pennis (Latin – "wings"); having iridescent wings

Label 2 Tetragonula iridipennis (Smith, 1854)

Image Ref:
Stingless bees (Hymenoptera: Apidae: Meliponini) of the Indian subcontinent: Diversity, taxonomy and current status of knowledge
Claus Rasmussen 2013 http://dx.doi.org/10.11646/zootaxa.3647.3.1

Taxonomic Notes On Stingless Bee *Trigona* (*Tetragonula*) *Iridipennis* Smith (Hymenoptera: Apidae) From India
K. Vijayakumar & R. Jeyaraaj 2014 http://dx.doi.org/10.11609/JoTT.o3773.6480-4

Akademi Kelulut Malaysia

Illustration 7 Tetragonula iridipennis (Smith, 1854)

6.9 *Tetragonula melina* (Gribodo, 1893)

Trigona melina **Gribodo** 1893: 262-263, 264: **Syntypes** (MSNG, three workers; unknown depository, two workers): Two syntypes are labelled "Bandjarmas" (?=Bandjarmasin, near Liangtelan), while the third syntype carry no locality label (F. Penati, pers. comm.) (taxonomy); **Type locality: MALAYSIA** "Liangtelan (Borneo)" (2 workers); "Perak (Malacca)" (3 workers);

melina, *Tetragonula* (Gribodo, 1893)[#] (Rasmussen, 2008) = Latin-From melinus ("colour of honey")

Label 3 Tetragonula melina (Gribodo, 1893)

https://data.nhm.ac.uk/dataset/56e711e6-c847-4f99-915a-6894bb5c5dea/resource/05ff2255-c38a-40c9-b657-4ccb55ab2feb?view_id=6ba121d1-da26-4ee1-81fa-da11e68f68e&filters=country%3AMalaysia%7CcollectionCode%3Abmnh%28e%29%7Cfamily%3Aapidae%7Cgenus%3Atetragonula&_genus_limit=0

Tetragonula melina (Gribodo, 1893)

The thorax and likewise at least the legs and abdomen honey-colored, entire body honey-testacesous in color except that the flagellum is as a rule dark above..............Tetragonula melina (Gribodo 1893) *BORNEO; THAILAND; MALAYSIA; INDONESIA- SUMATRA
BL 5 nim, WL, 4.7- 5.0 mm (4.5- 5 and 5.5 mm in Schwarz 1939),
HW 2.1 2.2 mm, WL2 1.5-1.6 mm, HTL 2.1-2.2.

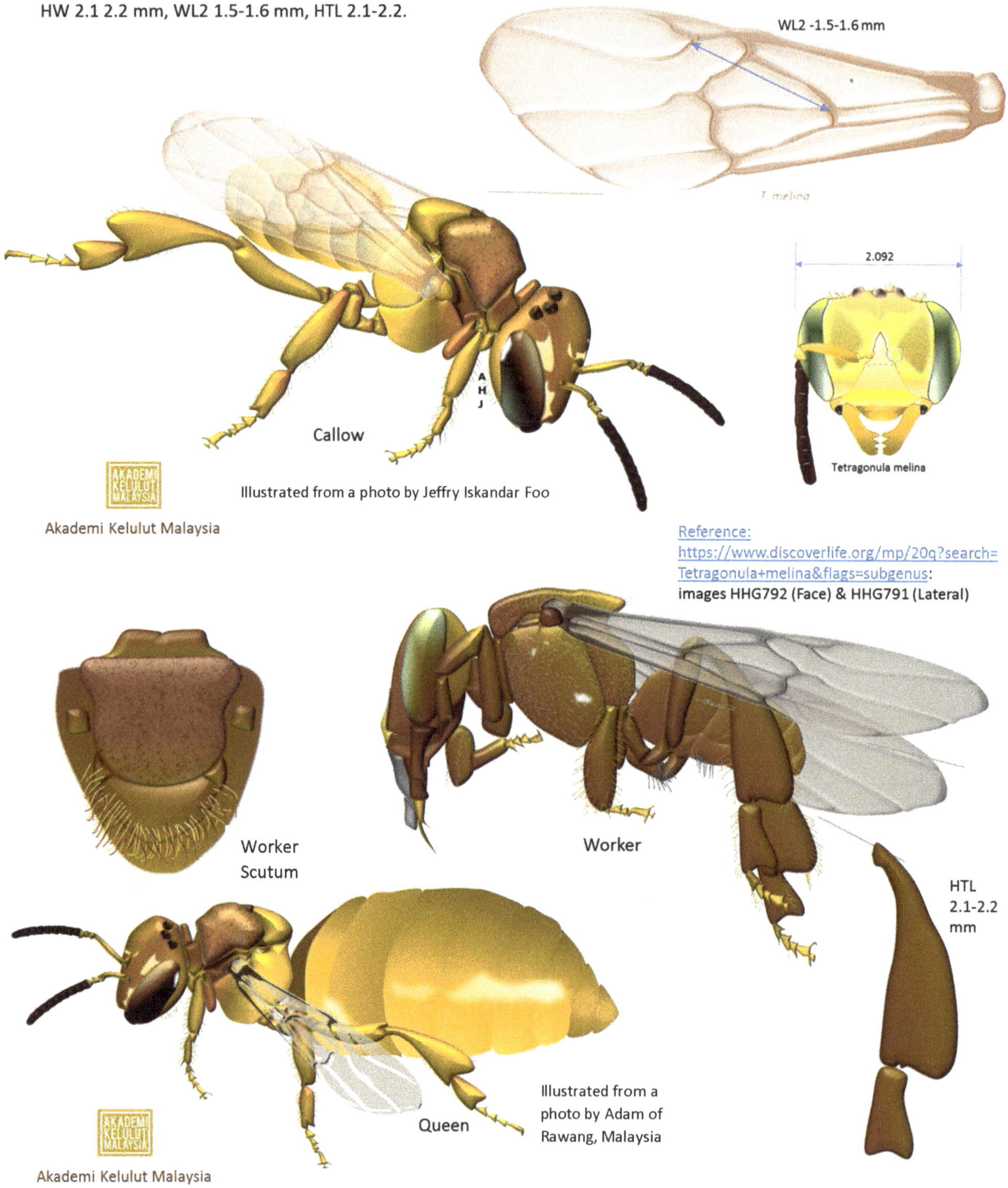

Reference: https://www.discoverlife.org/mp/20q?search=Tetragonula+melina&flags=subgenus: images HHG792 (Face) & HHG791 (Lateral)

Illustration 8 Tetragonula melina (Gribodo, 1893)

6.10 *Tetragonilla collina* (Smith, 1857)

Yoshikawa et al. 1969: 173 (distribution); MALAYSIA, George Town, Penang;

collīna, *Tetragonilla* (Smith, 1857) [Typ. loc.: Malaysia "Malacca (**Mount** Ophir)"][#] (Rasmussen, 2008) = From Latin - **collīnus** m (*feminine* **collīna**, *neuter* **collīnum**); *first/second declension* - Of a hill; growing on a mountain. **Mount Ophir**, or more commonly known by its Malay name, **Gunung Ledang**, is a mountain situated in the Gunung Ledang National Park located in Ledang District (northwestern Johor), Malaysia. The summit located between the border of Muar and Malacca

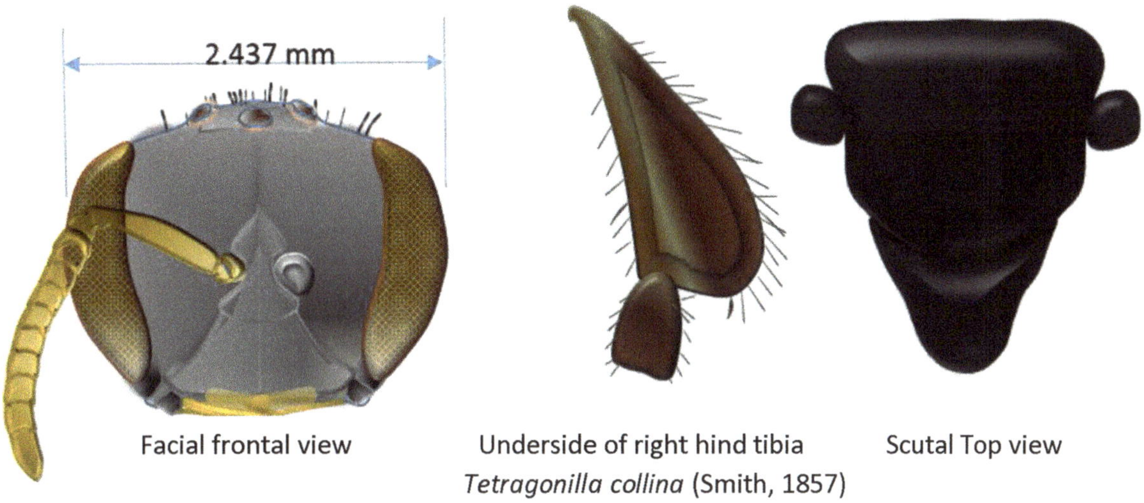

Illustration 9 Tetragonilla collina (Smith, 1857)

6.11 *Lepidotrigona terminata* (Smith, F., 1878)

Syn. *Trigona fulvomarginata* **Cockerell** 1919c: 78: **Holotype** (BMNH 17b.1083) (key to species); **Type locality: MALAYSIA** "Penang" (worker);

Eupatorium makinoi — *Mangifera indica* (Mango) — *Mangifera foetida* (Machang)

Figure 43 Lepidotrigona terminata preferred forage

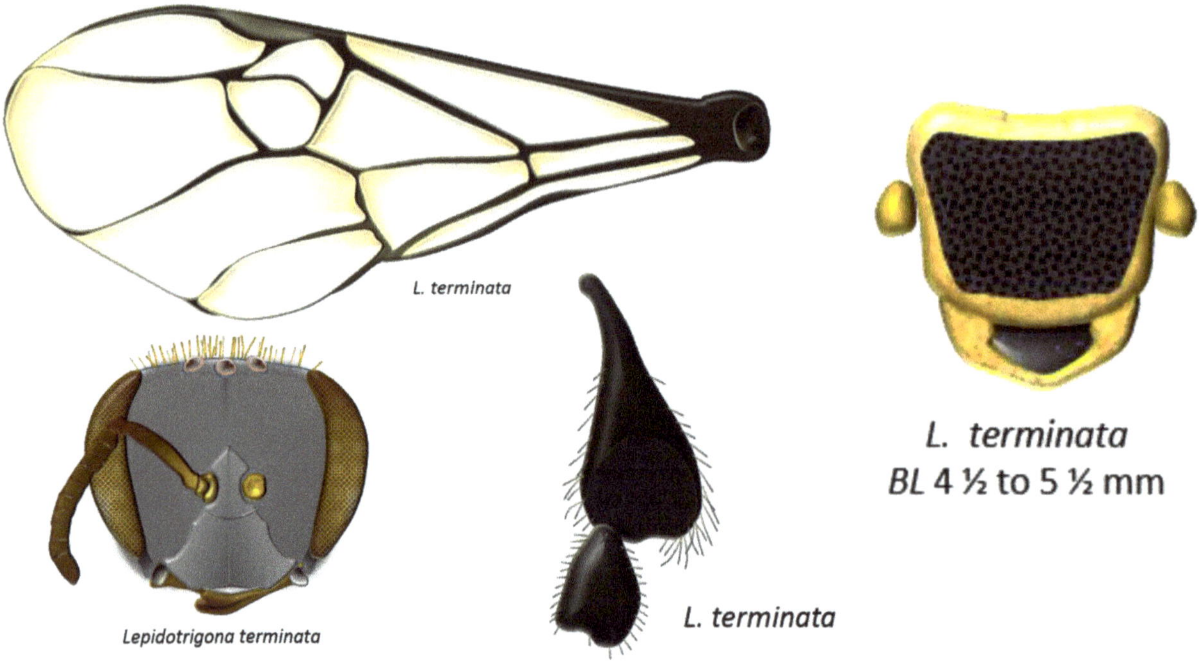

Illustration 10 Lepidotrigona terminata (Smith, F., 1878)

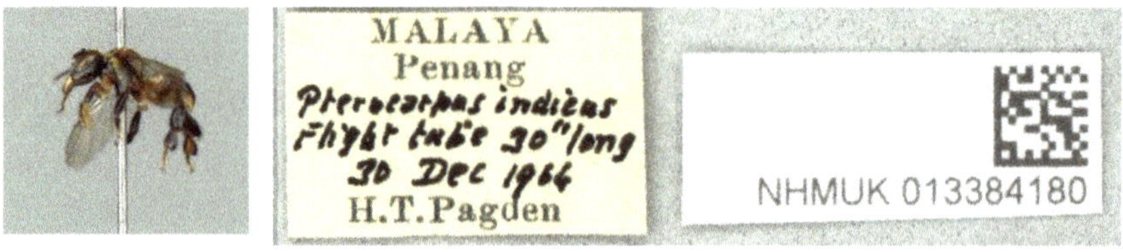

Label 4 Lepidotrigona terminata (Smith, F., 1878)

6.12 *Heterotrigona bakeri* (Cockerell, 1919)

Trigona bakeri **Cockerell** 1919c: 78, 79: **Holotype** (USNM 29468, worker): examined, "Island of / Penang / Baker", "9069", "Type No. / 29468" (head glued to label), "Trigona / bakeri / Ckll TYPE" (comparative notes, key to species); **Type locality: MALAYSIA** "Penang Island (Baker 9068)" (worker); **Cockerell** 1930a: 157 (distribution, floral record, variation)

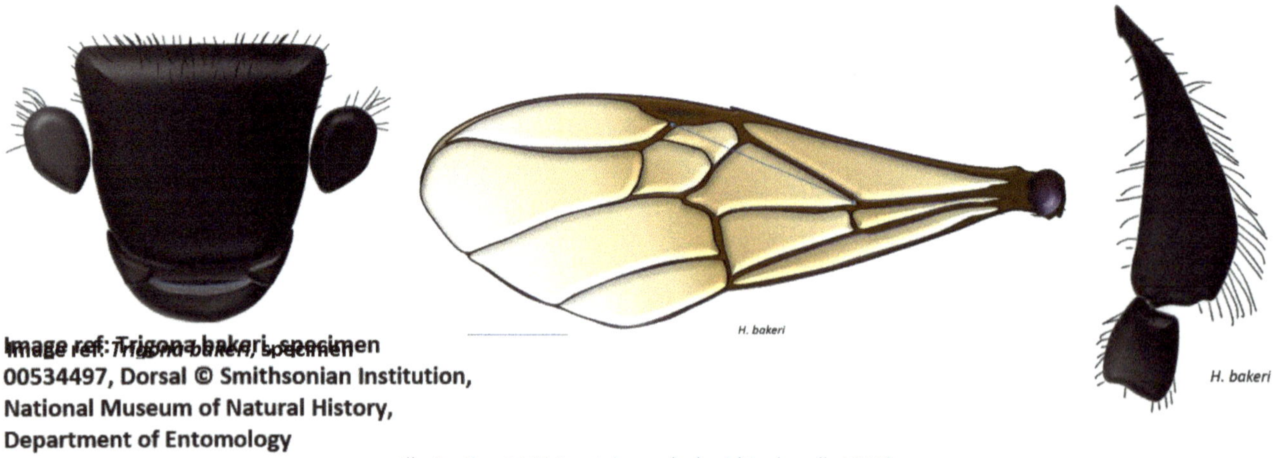

Illustration 11 Heterotrigona bakeri (Cockerell, 1919)

The following couplet is based on an examination of H. itama and descriptions in the literature (Cockerell 1919c, pp. 78-79; Schwarz 1939.).

(a) Scutellum is conspicuously hairy. ....*Heterotrigona itama* (Cockerell 1918)*

(See Schwarz 1939, p. 96).

THAILAND; MALAYSIA- PENINSULA, PENANG IS.; SINGAPORE; INDONESIA- JAVA, SUMATRA; BORNEO

(b) Scutellum bare. ....*Heterotrigona bakeri* (Cockerell 1919)

MALAYSIA- **PENANG IS**.; CAMBODIA

Source: Images above courtesy of Saiful Baha of Kuantan.

Figure 44 Heterotrigona bakeri (Cockerell, 1919) hairless scutellum

There is variability in the obscuration of the wing. The description of itama (1919, p. 79) reads "with dilute fuliginous wings", and this phrase also applied in the description of the wings of *breviceps* (1919A, p. 244), but Cockerell (1919A, p. 241) designated a variety of *itama* from the Island of Penang that had wings "somewhat darker than in typical itama".

Following are excerpts of Schwarz's descriptions in 1939: *"Trigona bakeri*, which Cockerell separated from itama, based on its bare scutellum and smoother and more polished mesothorax, is in my estimation hardly to be distinguished structurally from *itama*. The greater shininess of the mesonotum of bakeri seems to be due not to a more polished condition of the chitin but to the absence of the fine, appressed down that covers this area in itama. *Trigona bakeri* is a little less hairy on its mesothorax and over the head in front. However, these distinctions are bridged by yet a third specimen example from the type locality of bakeri, Island of Penang, that bears the identical field data. This specimen was withheld by Baker when the specimen on which Cockerell erected bakeri was submitted to Cockerell and was subsequently acquired by the U. S. National Museum. It shares with *itama* the downy appressed hairs on the mesonotum and hence the more subdued sheen; its scutellum is hairy like that of *itama*, not mostly bare like that of the type of *bakeri*. It would seem doubtful, therefore, whether *bakeri* represents more than an individual variation from *itama*." (Schwarz 1939)

"The type of *itama*, like the type of *bakeri*, is in the U. S. National Museum; the type of *breviceps* is in the British Museum. I have not had a chance, therefore, to compare the former with the last mentioned. The homotypes I had made of *itama*, however, seem hardly separable from *breviceps*, both the length and width of head and the malar space measured on a micrometre scale being substantially in accord with the corresponding parts of *breviceps*. I have tried to indicate the hair's breadth difference in the key that may exist, using in this connection specimens of *itama* from Singapore, the type locality of *itama* in the British Museum collection. Still, it is difficult to evaluate the difference with accuracy." (Schwarz 1939)

Chapter 7

7.0 Historical records of Highland stingless bees in Malaysia

From Professor T. D. A. Cockerell, I have received the following statement, drawn up by H. M. Pendlebury, explaining the relatively greater abundance of stingless bees in Malaysia as compared with the areas beyond: "I suppose that the reason why Trigona are so numerous in Siam, Malaya, and Borneo, and so poor outside these limits is possibly due to the greater number of resin-secreting trees belonging to Dipterocarpaceae[51] found in Malaysia; the stable conditions regarding humidity and temperature may be contributory factors also." (Excerpt from Schwarz 1937 – endnote pg. 284)

7.0.1 Records of *itama* from Schwarz 1939

State of Perak:

- Maxwell's Hill, 3570 feet, April 9, 1904;
- Batang Padang, Jor Camp,2000 feet, Aug. 21, 1922, and Sept. 6, 1922 (E. Seimund);
- Bukit Merah, Feb. 23, 1930 (H. T. Pagden);
- Taiping Hill, 500-1500 feet, Dec. 11, 1923 (M. R. Henderson), 3700feet, Sept. 12, 1931 (H. T. Pagden);
- Larut Hills, G. Ijam, 3750 feet and 4750 feet,- Feb. 10, 1932 (H. M. Pendlebury);
- Larut Hills, 4500 feet, Feb. 18, 1932, and 3700-4500 feet, Feb. 24, 1932 (H. M. Pendlebury).

State of Selangor:

- Batu Caves, Oct. 25, 1906 (Agri. Dept.), Sept. 19, 1921, and Jan. 2, 1925 (H. M. Pendlebury); nr. Batu Caves, Aug. 22, 1926 (C. Dover);
- Genting Bidai, 2000 (C. B. Kloss);
- Bukit Kutu[52], 3500 feet, April 16-19, 1926, Sept. 6, 1929, and March 14, 1931 (H. M. Pendlebury) and 3485 feet, Jan. 30, 1930 (H. T. Pagden);
- the foot of Bukit Kutu, on *Melastoma*, Jan. 19, 1936 (H. T. Pagden).

State of Pahang: Sungei Tahan, Nov. 22, 1921 (H. M. Pendlebury);

- Kuala Tahan, Nov. 22, 1921 (H. M. Pendlebury);
- Lubok Tamang, 3500 feet, June 15-16, 1923 (H. M. Pendlebury), and. 3600 feet, May 25, 1931 (H. T. Pagden);
- Kuala Lipis, May 29, 1926 (H. M. Pendlebury);
- Jerantut-Batu Balai, March 18-19, 1927 (E. Seimund);
- Fraser's Hill, 4000 feet, Jan. 25, 1929, and Feb. 2, 1929 (H. M. Pendlebury), Sept. 30, 1929 (N. C. E. Miller), May 1935 (G. H. Lowe), and 4200 feet, June 18, 1931, and July 15, 1936 (H. M. Pendlebury);
- Cameron Highlands, 3600 feet, May 26, 1931 (H. T. Pagden), 4000-5000 feet, June 10, 1935 (H. M. Pendlebury), 4800 feet, June 22, 1935 (H. M. Pendlebury), 4500-4800 feet, June 23, 1935 (H. M. Pendlebury);

---

[51] Characters of Dipterocarpaceae: Habit: Gigantic trees rarely shrubs. Inflorescence: Paniculate terminal or axillary spikes or racemes. Calyx: Sepals 5, unequal, polysepalous persistent and wing-like in fruit (*Shorea*), imbricate or valvate. The plants of Dipterocarpaceae are of great value for resins and timber, oils, camphor and turpentine.

[52] Mr. Pagden states that the meaning of Bukit Kutu is literally Hill of the Louse. It is one of the older hill stations and one of the most interesting collecting areas he has visited.

- Cameron Highlands, Tanah Rata to Padang, 4800 feet, May 30 to June 2, 1931 (H. M. Pendlebury); Tanah Rata, 4600 feet, May 25, 1931 (H. T. Pagden); 4800 feet, June 2, 1931 (H. M. Pendlebury).

State of Selangor-State of Pahang:

- border The Gap, 2700 feet, May 25, 1926 (H. M. Pendlebury);
- Genting Sempah Hill Stream, Sept. 11, 1926 (C. Dover).

Island of Singapore:

- Singapore (Baker), including a male; Bt. Timah, July 11, 1911;
- Mandai, Dec. 29, 1922 (H. C. Abraham).

West Coast:

- Langkawi Island, April 14-21, 1928 (H. M. Pendlebury).

N. Borneo. -

- Bettotan, nr. Sandakan, July 23, 1927 (C. B. Kloss), Aug. 24, 1927 (H. M. Pendlebury);
- nr. Kinabalu, Tenompok Pass, 4700 feet, March 18, 1929 (H. M. Pendlebury);
- Mt. Kinabalu, Kenokok, 3300 feet, April 23, 1929 (H. M. Pendlebury);
- nr. Kinabalu, Kabayau, May 9-11, 1929 (H. M. Pendlebury).

7.0.2 Records of *erythrogastra* from Schwarz 1939

State of Kedah:

- nr. Jitra, Catchment Area, April 9, 1928 (H. M. Pendlebury).

State of Perak:

- Gunong Kledang, 2646 feet, Nov. 1916;
- Batang Padang, Jor Camp, 2000 feet, Aug. 21, 1922 (E. Seimund), 1800 feet, March 13, 1924, and Jan. 22, 1925 (H. M. Pendlebury).

State of Selangor:

- Ulu Gombak, 16th mile from Kuala Lumpur, March 23, 1928; 12th mile from Kuala Lumpur, July 28, 1929 (H. T. Pagden);
- Bukit Kutu, 3500 feet, Sept. 6, 1929 (H. M. Pendlebury);
- 3485 feet, Jan. 1, 1930 (H. T. Pagden), foot of the hill on Melastoma, Jan. 19, 1936 (H. T. Pagden).

State of Pahang:

- Jerantut-Kuantan Road, 101st mile, Aug. 30, 1926 (H. M. Pendlebury);
- Jerantut-Batu Balai, March 18, 1927 (E. Seimund);
- Kuala Lipis, May 29, 1928 (N. C. E. Miller);
- Ulu Sungai Triang, June 1930 (D. S. Inglis).

State of Selangor- State of Pahang:

- Genting Sempah Hill Stream, Sept. 11, 1926 (C. Dover).

N. Borneo.-

- Samawang, nr. Sandakan, jungle, July 1927 (C. B. Kloss and H. M. Pendlebury);
- Bettotan, nr. Sandakan, Aug. 2-19, 1927 (C. B. Kloss and H. M. Pendlebury).

7.0.3 Records of *canifrons* from Schwarz 1939

State of Selangor:

- Bukit Cherakah, July 18, 1921 (H. M. Pendlebury);
- 12th mile Hot Spring Klang Gates, Jan. 9, 1925 (H. M. Pendlebury);
- 12th mile Ulu Gombak, Jan. 14, 1925 (H. M. Pendlebury);
- Bukit Kutu, 200 feet, Sept. 21, 1932 (H. M. Pendlebury), 3300-3500, Sept. 23, 1932 (H. M. Pendlebury), 3485 feet, Jan. 30, 1930 (H. T. Pagden);

- Klang Gates Reservoir, Sept. 29, 1930;
- Kanching, July 26, 1936 (H. T. Pagden).

State of Pahang:

- Tembeling, Nov. 19, 1921 (H. M. Pendlebury);
- Kuala Tahan, Nov. 25, 1921, and Nov. 20, 1922 (H. M. Pendlebury);
- Sungei Tahan, Nov. 22, 1922 (H. M. Pendlebury);
- Kuala Lipis, May 29, 1926 (H. M. Pendlebury);
- Batu Balai-Jerantut, March 19, 1927 (E. Seimund);
- Chintamani, Aug. 21, 1935.

State of Johore:

- Sungai Palong, May 15- 23, 1930 (D. S. Inglis).

Island of Singapore:

- Bukit Timah, July 26, 1911; Singapore (Baker).

N. Borneo.-

- Bettotan, nr. Sandakan, Aug. 19, 1927 (C. B. Kloss and H. M. Pendlebury);
- Sandakan (Baker).

7.0.4 Records of *fimbriata* from Schwarz 1939

State of Kedah:

- Gunong Jerai or Kedah Peak, 3300-3950 feet, March 11-14, 1928, and 3950 feet, March 9, 1928 (H. M. Pendlebury), 3978 feet, April 20, 1930, and 4000 feet, Dec. 30, 1931 (H. T. Pagden).

State of Perak:

- Larut Hills, 3700-4500 feet, Feb. 14, 1932, and 4500 feet, Feb. 21, 1932 (H. M. Pendlebury).

State of Selangor:

- Bukit Kutu, 3300 feet, Sept. 26, 1932 (H. M. Pendlebury), 3300-3500 feet, March 15-20, 1931 (H. M. Pendlebury), 3400 feet, March 13, 1931 (H. M. Pendlebury), 3400-3500 feet, March 17, 1931 (H. M. Pendlebury), 3500 feet, April 10-19, 1926, Sept. 7-11, 1929, and March 14-16, 1931 (H. M. Pendlebury);
- Ampang, June 14, 1936 (H. M. Pendlebury).

State of Pahang:

- Kuala Teku, Aug.-Nov. 1920 (H. M. Pendlebury);
- Gunong Tahan, Wray's Camp, Nov. 1920 (J. Bangga);
- Beserah, May 31, 1926 (H. M. Pendlebury);
- Batu Balai, Jerantut, March 19, 1927 (E. Seimund);
- Kuala Lipis, May 29, 1928 (N. C. E. Miller);
- Fraser's Hill, 4000 feet, July 20, 1936; 4200 feet, July 19, 1936 (H. M. Pendlebury).

State of Selangor-State of Pahang:

- The Gap, 2700 feet, Jan. 1, 1915.

East Coast:

- Tioman Island, June-July 1916 (H. C. R. and C. B. K.).

N. Borneo.-

- Samawang, nr. Sandakan, jungle, July 25, 1927 (C. B. Kloss and H. M. Pendlebury);
- Bettotan, nr. Sandakan, Aug. 3-13, 1927 (C. B. Kloss and H. M. Pendlebury).

7.0.5 Records of *thoracica* from Schwarz 1939

State of K6dah:

- Alor Star, Gunong Keriang, April 1, 1928 (H. M. Pendlebury);

- nr. Jitra, Catchment Area, April 7, 1928 (H. M. Pendlebury).

State of Perak:

- Maxwell's Hill, Aug. 16, 1908;
- Batang Padang, Jor Camp, 1800 feet, Jan. 22, 1925 (H. M. Pendlebury);
- Kuala Kangsar, on flowers of *Tephrosia candida*, May 16, 1928 (H. T. Pagden);
- Maxwell's Garden, Taiping Hill, 3300 feet, Sept. 12, 1931 (H. T. Pagden);
- Taiping Hill, 3700 feet, Sept. 11-13, 1931 (H. T. Pagden);
- Larut Hills, 3000-3700 feet, Feb. 16, 1932 (H. M. Pendlebury).

State of Selangor:

- Batu Caves, Dec. 6, 1923, and Jan. 2, 1925 (H. M. Pendlebury);
- Kuala Lumpur, Welch Hill, Oct. 8, 1922 (H. M. Pendlebury);
- Dusun Tua, July 30, 1921 (H. M. Pendlebury);
- Ulu Klang District, taken flying around a hot sulphur spring, Aug. 25, 1929 (H. T. Pagden);
- Bukit Kutu, 3485 feet, Jan. 30, 1930 (H. T. Pagden), 3300-3500 feet, March 15, 1931 (H. M. Pendlebury);

State of Pahang:

- Gunong Tahan, Wray's Camp, 3300 feet, Nov. 1920; Kuala Tahan, Nov. 25, 1921 (H. M. Pendlebury);
- Kuala Teku, in "Belukar" (Shrubbery), 500 feet, Dec. 7, 1921 (H. M. Pendlebury);
- Batu Balai-Jerantut, March 19, 1927 (E. Seimund);
- Kuala Lipis, May 29, 1928 (N. C. E. Miller);
- Fraser's Hill, 4000 feet, Jan. 27-29, 1929, and Feb. 2, 1929 (H. M. Pendlebury), Sept. 30, 1929
- (N. C. E. Miller), May 1935 (G. H. Lowe), 4200 feet, June 18, 1931 (H. M. Pendlebury);
- nr. Karak, Chintamani Jungle, Aug. 22, 1935.

Island of Singapore:

- Bt. Timah, June 30, 1911;
- Singapore Botanic Gardens, July 11, 1911.

N. Borneo.-

- Bettotan, nr. Sandakan, July 28, 1927 (C. B. Kloss and H. M. Pendlebury);
- nr. Kinabalu, Kabayau, 600 feet, May 9, 1929 (H. M. Pendlebury).

7.0.6 Records of *iridipennis* from Schwarz 1939

State of Kedah:

- Bukit Panchor, July 10, 1929 (H. T. Pagden), Feb. 21, 1930 (R. P. Kelly).

State of Pahang:

- Tahan River, left bank, 300 feet, Nov. 29, 1921 (H. M. Pendlebury);
- Kuala Teku, 500 feet, Dec. 5, 1921 (H. M. Pendlebury), 500-1000 feet, Nov. 25, 1922 (H. M.
- Pendlebury);
- Kuala Tahan, 300 feet, Nov. 22, 1921 (H. M. Pendlebury); 10 3rd mile Jerantut-Kuantan Road, June 2, 1926 (H. M. Pendlebury);
- Sungei Tahan, Nov. 22, 1922 (H. M. Pendlebury).

West Coast:

- Langkawi Is., April 15-22, 1928 (H. M. Pendlebury);
- Pulau Jarak, April 6-12, 1932 (E. Seimund).

East Coast:

- Pulau Perhentian, June 14, 1926 (C. B. Kloss).

N. Borneo.-

- Samawang, nr. Sandakan, jungle, July 12, 1927, and July 17, 1927 (C. B. Kloss and H. M. Pendlebury);
- Bettotan, nr. Sandakan, July 22-Aug. 24, 1927, Aug. 13, 1927, males (C. B. Kloss and H. M. Pendlebury).

7.0.7 Records of *valdezi* from Schwarz 1939

State of Kedah:

- Gunong Jerai or Kedah Peak, 3000 feet, March 7-8, 1928 (H. M. Pendlebury), 3000-3500 feet, March 15, 1928 (H. M. Pendlebury), 3300 feet, March 22, 1928 (H. M. Pendlebury), 4000 feet, Dec. 30, 1931 (H. T. Pagden).

State of Perak:

- Batang Padang, Jor Camp, 1800 feet, Jan. 21, 1925 (H. M. Pendlebury); Batang Padang, Tapah, March 7, 1925 (H. M. Pendlebury).

State of Selangor:

- Bukit Kutu, 3500 feet, Sept. 13-14, 1929, and March 14, 1931 (H. M. Pendlebury);
- Tanjong Mas Est., Sepang, Jan. 20, 1928 (H. T. Pagden).

State of Pahang:

- Kuala Tahan, 300 feet, Nov. 21, 1921 (H. M. Pendlebury);
- Sungei Tahan, Nov. 22, 1922 (H. M. Pendlebury);
- Tanah Rata, 5000 feet, May 24, 1931 (H. T. Pagden).

State of Selangor- State of Pahang:

- nr. Genting Sempah Hill Stream, Sept. 11, 1926 (C. Dover).

East Coast:

- Pulau Perhentian, June 14, 1926 (C. B. Kloss).

West Coast:

- Langkawi Is., May 1, 1928.

N. Borneo.-Sandakan (Baker);

- Bettotan, nr. Sandakan, April 3, 1927 (C. B. Kloss and H. M. Pendlebury);
- Samawang, nr. Sandakan, July 16, 1927 (C. B. Kloss and H. M. Pendlebury);
- Kudat, Sept. 10-15, 1927 (C. B. Kloss and H. M. Pendlebury);
- Mt. Kinabalu, Kenokok, 3300 feet, April 23, 1929 (H. M. Pendlebury);
- nr. Kinabalu, Kabayau, 600 feet, May 7, 1929 (H. M. Pendlebury).
- Labuan Is., Sept. 23, 1927.

7.0.8 Records of *geissleri* from Schwarz 1939

State of Kedah:

- Kedah Peak or Gunong Jerai, 3000 feet, March 8, 1928 (H. M. Pendlebury), 3300 feet, March 12, 1928, male, and March 22, 1928 (H. M. Pendlebury), 3800 feet, Dec. 30, 1931 (H. T. Pagden).

State of Perak:

- Taiping Hill, 3700 feet, Sept. 11, 1931 (H. T. Pagden);
- Larut Hills, 4500 feet, Feb. 20-22, 1932 (H. M. Pendlebury).

State of Selangor:

- Bukit Kutu, 500-1000 feet, April 12, 1926 (H. M. Pendlebury), 3300-3500 feet, March 22, 1931 (H. M. Pendlebury), 3500 feet, Sept. 7, 1928, male, and Sept. 11-13, 1929 (H. M. Pendlebury);
- Kuala Lumpur, on flowers of a creeper, May 18, 1926 (H. M. Pendlebury), Entom. Lab., March 27, 1928 (H. T. Pagden).

State of Pahang:

- Wray's Camp, 300 feet, Dec. 9, 1921 (H. M. Pendlebury), 3300 feet, Dec. 11, 1921 (H. M. Pendlebury);
- Gunong Tahan Padang, 4900 feet, Dec. 17, 1922 (H. M. Pendlebury), 5500 feet, Dec. 7, 1922, Dec. 11, 1922, at a light, Dec. 23, 1922, Jan. 18, 1923, and Jan. 22, 1923 (H. M. Pendlebury);
- Cameron Highlands, Rhododendron Hill, 5200 feet, June 19-22, 1923 (H. M. Pendlebury); Tanah Rata to Padang, May 30, 1931 (H. M. Pendlebury); Tanah Rata, 4800 feet, June 2, 1931 (H. M. Pendlebury); G. Berembun summit, 6050 feet, March 14, 1924 (H. M. Pendlebury);4000-5000 feet, June 6, 1935, and June 17, 1935 (H. M. Pendlebury), 5000 feet, June 11, 1935 (H. M. Pendlebury);
- Fraser's Hill, 3500-4500 feet, Aug. 12, 1923 (H. M. Pendlebury), 4000 feet, Jan 27-29, 1929, May 28, 1932, males, and Oct. 25, 1933 (H. M. Pendlebury), May 1935 (G. H. Lowe), 4200 feet, May 26, 1926, June 27, 1931, July 6, 1931, and July 15, 1936 (H. M. Pendlebury).

Singapore Island:

- Mandai, 50 feet, March 18, 1922 (H. C. Abraham).

7.0.9 Records of *atripes* from Schwarz 1939

State of Kedah:

- nr. Jitra, Catchment Area, April 4-7, 1928 (H. M. Pendlebury);
- Bukit Panchor, June 4, 1930 (H. T. Pagden).

State of Perak:

- Jor Camp, 2000 feet, Aug. 25, 1922, and Sept. 6, 1922 (E. Seimund);
- Batang Padang, Jor Camp, 1800 feet, May 28, 1923, and March 10, 1924 (H. M. Pendlebury), 2000 feet, June 3, 1923 (H. M. Pendlebury);
- Taiping, Dec. 3, 1923 (M. R. Henderson).

State of Selangor:

- 17th-mile Kanching, Oct. 22, 1922 (H. M. Pendlebury);
- Genting Bidai, 2000 feet (C. B. Kloss); Serdang, on flowers of Citrus limon, April 25, 1928 (H. T. Pagden);
- Ulu Gombak, 12th mile from Kuala Lumpur, June 24, 1928 (H. T. Pagden); Ulu Langat, Aug. 19, 1928 (H. T. Pagden);
- Bukit Kutu, 200 feet, Sept. 21, 1932 (H. M. Pendlebury), 3300 feet, March 12, 1931, and Sept. 23, 1932 (H. M. Pendlebury), 3300-3500 feet, March 18-27, 1931 (H. M. Pendlebury), 3485 feet, Jan. 30, 1930 (H. T. Pagden), 3500 feet, April 17, 1926, and Sept. 6-12, 1929 (H. M. Pendlebury).

State of Pahang:

- Kuala Tahan, 300 feet, Nov. 21, 1921, and Nov. 22, 1921 (H. M. Pendlebury);
- Lubok Tamang, 3500 feet, June 13, 1923, Oct. 20, 1923, and March 28, 1924 (H. M. Pendlebury), 4000 feet, Sept. 7, 1922 (E. Seimund);
- Fraser's Hill, 4000 feet, May 1935 (G. H. Lowe);
- Cameron Highlands, 4800 feet, June 24, 1935 (H. M. Pendlebury).

Island of Penang (Baker).

7.0.10 Records of *collina* from Schwarz 1939

State of Kedah:

- Kedah Peak or Gunong Jerai, 2000-2500 feet, March 7, 1928 (H. M. Pendlebury), 3000 feet, March 8, 1928 (H. M. Pendlebury); 3300-3950 feet, March 9, 1928 (H. M. Pendlebury).

State of Perak:

- Batang Padang, Jor Camp, June 1923 (F. N. C.), Jor Camp, 1800 feet, Jan. 21, 1925 (H. M. Pendlebury), Jor Camp, 2000 feet, Aug. 21, 1922 (H. M. Pendlebury);

- Batang Padang, Jor Camp-Tamang, 7th mile, March 10, 1924 (H. M. Pendlebury).

State of Selangor:

- Genting Sempah, 2000 feet (C. B. Kloss); Genting Bidai, 2000 feet (C. B. Moss);
- Bukit Kutu, foot of the hill to 500 feet, Sept. 5, 1929 (H. M. Pendlebury), 3500 feet, Sept. 6, 1929 (H. M. Pendlebury);

State of Pahang:

- Kuala Teku, in Belukar, 500 feet, Dec. 7, 1921 (H. M. Pendlebury);
- Fraser's Hill, 4000 feet, Jan. 27, 1929, and May 31, 1932 (H. M. Pendlebury), and May 1935 (G. H. Lowe), 4200 feet, July 17, 1936 (H. M. Pendlebury);
- Ulu Sangai Triang, June 1930 (D. S. Inglis).

State of Selangor-State of Pahang:

- nr. Genting Sempah Hill Stream, Sept. 11, 1926 (C. Dover).

State of Johore:

- Sungai Palong, May 15-23, 1930 (D. S. Inglis).

Island of Penang (Baker).

N. Borneo.-

- Samawang, nr. Sandakan, jungle, July 6-9, 1927 (C. B. Kloss and H. M. Pendlebury);
- Bettotan, nr. Sandakan, Aug. 4-22, 1927 (C. B. Kloss and H. M. Pendlebury).

7.0.11 Records of *fuscibasis* from Schwarz 1939

State of Selangor:

- Bukit Kutu, 3500 feet, April 18, 1926, and Sept. 6, 1929 (H. M. Pendlebury).

State of Pahang:

- Gunong Tahan, Wray's Camp, Nov. 1920 (H. Bangga);
- Sungei Tahan, Nov. 22, 1922 (H. M. Pendlebury).

States of Selangor-Pahang:

- Genting Sempah Hill Stream, Sept. 11, 1926 (C. Dover).

State of Johore:

- Mt. Ophir (type locality of variety *collina*), Aug. 11, 1905.

N. Borneo.-

- Samawang, nr. Sandakan, July 9,1927 (C. B. Kloss and H. M. Pendlebury);
- Bettotan, nr. Sandakan, July 31-Aug. 23, 1927 (C. B. Kloss and H. M. Pendlebury);
- Kudat, Sept. 3, 1927 (C. B. Kloss and H. M. Pendlebury).

7.0.12 Records of *rufibasalis* from Schwarz 1939

State of Pahang:

- Kuala Tahan, 300 feet, Nov. 21, 1921 (H. M. Pendlebury).

N. Borneo.-

- Samawang, nr. Sandakan, jungle, July 9-12, 1927 (C. B. Kloss and H. M. Pendlebury);
- Bettotan, nr. Sandakan, July 24, Aug. 3, Aug. 19, and Aug. 20, 1927 (C. B. Kloss and H. M. Pendlebury).

7.0.13 Records of *apicalis* from Schwarz 1939

State of Kedah:

- Kedah Peak, 3300 feet, March 20, 1928 (H. M. Pendlebury);
- nr. Jitra, Catchment Area, April 7-11, 1928 (H. M. Pendlebury).

State of Perak:

- Batang Padang, Jor Camp, 1800 feet, June 4, 1923, and Oct. 10, 1923 (H. M. Pendlebury), 2000 feet, Sept. 16, 1922 (E. Seimund);
- Larut Hills, 500-1000 feet, April 27, 1938 (H. M. Pendlebury).

State of Selangor:

- Bukit Cherakah, July 18, 1921 (H. M. Pendlebury);
- Gombak Valley, Oct. 11-22, 1921 (H. M. Pendlebury), 19th-22nd mile, April 6, 1924 (H. M. Pendlebury);
- 17th-mile Kanching, Oct. 22, 1922 (H. M. Pendlebury);
- Kanching, Dec. 28, 1930 (H. M. Pendlebury);
- Ampang, July 12, 1922 (H. M. Pendlebury);
- Ulu Gombak, March 4, 1928 (H. T. Pagden);
- Bukit Kutu, 3300 feet, March 12, 1931, and Oct. 2, 1932 (H. M. Pendlebury), 3400 feet, March 13, 1931 (H. M. Pendlebury), 3500 feet, April 19, 1926, and March 14, 1931 (H. M. Pendlebury), 3300-3500 feet, March 18-22, 1931 (H. M. Pendlebury).

State of Pahang:

- Kuala Tahan, Nov. 22, 1923 (H. M. Pendlebury).

West Coast:

- Langkawi Is., April 14-27, 1928 (H. M. Pendlebury).

East Coast:

- Tioman Is., June-July 1916 (H. C. R. and C. B. K.).

Singapore Island:

- Singapore, March 1907 (F. Muir);
- Bt. Timah, June 10, 1911;
- Botanic Garden, July 10, 1911, and June 25, 1926 (C. B. Kloss);
- Changi, Aug. 1911;
- Selitar, Aug. 1911.

N. Borneo.-

- Samawang, nr. Sandakan, jungle, July 1927 (C. B. Kloss and H. M. Pendlebury);
- Bettotan, nr. Sandakan, Aug. 15-22, 1927 (C. B. Kloss and H. M. Pendlebury);
- Kudat, Sept. 7-15, 1927 (C. B. Kloss and H. M. Pendlebury);
- Mt. Kinabalu, Kenokok, 3300 feet, April 22, 1929 (H. M. Pendlebury); nr. Kinabalu, Kabayau, 600 feet, May 8-10, 1929 (H. M. Pendlebury).

7.0.14 Records of *binghami* from Schwarz 1939

N. Borneo.-

- Samawang, nr. Sandakan, jungle, July 9-15, 1927 (C. B. Kloss and H. M. Pendlebury);
- Bettotan, nr. Sandakan, July 28-Aug. 6, 1927 (C. B. Kloss and H. M. Pendlebury);
- Kudat, Sept. 8, 1927 (C. B. Kloss and H. M. Pendlebury).

7.15 Records of *melanoleuca* from Schwarz 1939

State of Perlis:

- Bukit Jerneh, March 17, 1936 (H. M. Pendlebury).

State of Perak:

- Larut Hills, 3700-4000 feet, Feb. 12, 1932 (H. M. Pendlebury), 4500 feet, Feb. 20, 1932, and Feb. 22, 1932 (H. M. Pendlebury).

State of Selangor:

- Bukit Kutu, 3500 feet, Sept. 6, 1929, and March 16-21, 1931 (H. M. Pendlebury), 3000-3500 feet, March 1931, male (H. M. Pendlebury);
- Kanching Forest Reserve, Jan. 26, 1930 (H. T. Pagden).

State of Pahang:

- Fraser's Hill, 4000 feet, Nov. 22, 1936 (H. M. Pendlebury).

State of Selangor-State of Pahang:

- Semangko Pass, 2700 feet, March 1912.

N. Borneo.-

- Bettotan, nr. Sandakan, July 22, 1927 (C. B. Kloss and H. M. Pendlebury).

7.0.16 Records of *peninsularis* from Schwarz 1939

State of Selangor:

- Gombak Valley, Oct. 12, 1921, and Oct. 13, 1921 (H. M. Pendlebury).

State of Pahang:

- Wray's Hill, 900 feet, Dec. 9, 1921 (H. M. Pendlebury).

State of Selangor-State of Pahang:

- nr. Genting Sempah Hill Stream, Sept. 11, 1926 (C. Dover).

7.0.17 Records of *nitidiventris* from Schwarz 1939

State of Perak:

- Jor Camp, 2000 feet, Aug. 21, 1922 (E. Seimund);
- Batang Padang, Jor Camp-L. Tamang, 7th mile, March 10, 1924 (H. M. Pendlebury).

State of Selangor:

- Genting Bidai, 2000 feet (C. B. Kl ss);
- Bukit Kutu, 3485 feet, Jan. 30, 1930 (H. T. Pagden).

State of Pahang:

- Sungei Tahan, Nov. 22, 1922 (H. M. Pendlebury);
- Fraser's Hill, 4200 feet, June 2, 1936 (H. M. Pendlebury).

State of Selangor-State of Pahang:

- Genting Sempah Hill Stream, Sept. 11, 1926 (C. Dover).

7.0.18 Records of *trochanterica* from Schwarz 1939

State of Selangor:

- Bukit Kutu, 200 feet, Sept. 21, 1922 (H. M. Pendlebury).

State of Pahang.

- Gunong Tahan, Wray's Camp, Nov. 1920 (J. Bangga).

7.0.19 Records of *flavibasis* from Schwarz 1939

State of Perak: Perak, 1902;

- Jor Camp, 2000 feet, Aug. 21, 1922 (E. Seimund);
- Batang Padang, Jor Camp-L. Tamang, 7th mile, March 10, 1924 (H. M. Pendlebury).

State of Selangor:

- Ampang For. Reserve, Oct. 28, 1928 (H. M. Pendlebury).

State of Pahang:

- Fraser's Hill, Sept. 30, 1929 (N. C. E. Miller).

N. BORNEO. -

- Mt. Kinabalu, Kenokok, 3200 feet, April 23-24, 1929 (H. M. Pendlebury).

7.0.20 Records of *terminata* from Schwarz 1939

State of Perak:

- Batang Padang, Jor Camp-L. Tamang, 7th mile, March 10, 1924 (H. A. Pendlebury).

State of Selangor:

- Sungai Chongkak, Ulu Langat, on damp sand by the stream, Feb. 10, 1929 (H. T. Pagden);
- Bukit Kutu, 200 feet, Sept. 21, 1932 (H. M. Pendlebury);
- Bkt. Cherakah, Aug. 30, 1936 (H. T. Pagden).

State of Pahang:

- Kuala Teku Jungle, 500-1200 feet, Nov. 25, 1922 (H. M. Pendlebury).

State of Johore:

- Sungai Palong, May 15-23, 1930 (D. S. Inglis).

7.0.21 Records of *latebalteata* from Schwarz 1939

State of Perak:

- Jor Camp, 2000 feet, Aug. 21, 1922 (E. Seimund).

State of Selangor:

- Gombak Valley, Oct. 12, 1921 (H. M. Pendlebury).

State of Negri Sembilan:

- Port Dickson, Dec. 25-31 (H. M. Pendlebury).

N. Borneo.-

- Bettotan, nr. Sandakan, July 26, 1927, Aug. 16, 1927, Aug. 23, 1927 (C. B. Kloss and H. M. Pendlebury);
- Kudat, Sept. 16, 1927 (C. B. Kloss and H. M. Pendlebury);
- Mt. Kinabalu, Kiau, 3000 feet, March 25, 1929 (H. M. Pendlebury);
- nr. Kinabalu, Kabayau, 600 feet, May 11, 1929 (H. M. Pendlebury).

7.1 Cameron Highlands - a) *Tetragonula geissleri* (Cockerell 1918)

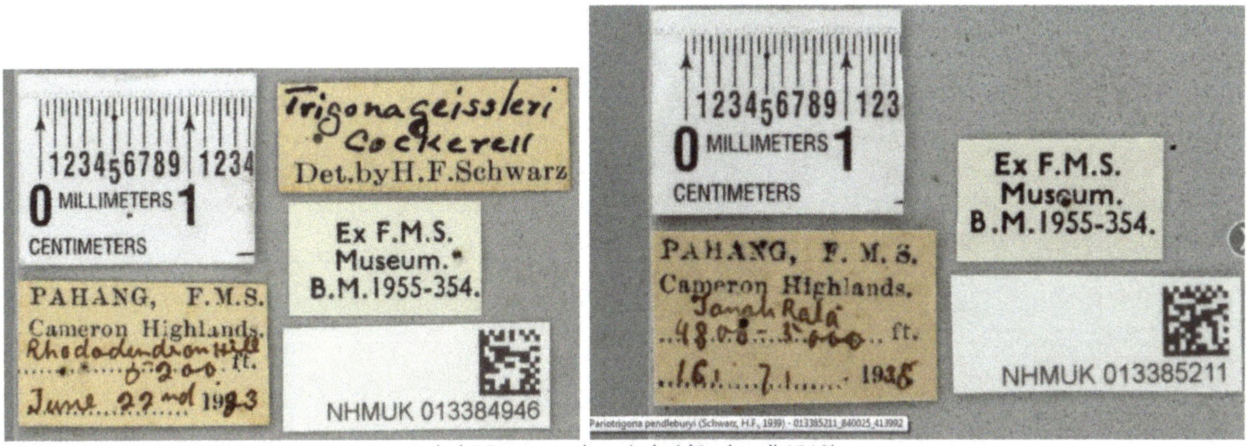

Label 5 Tetragonula geissleri (Cockerell 1918)

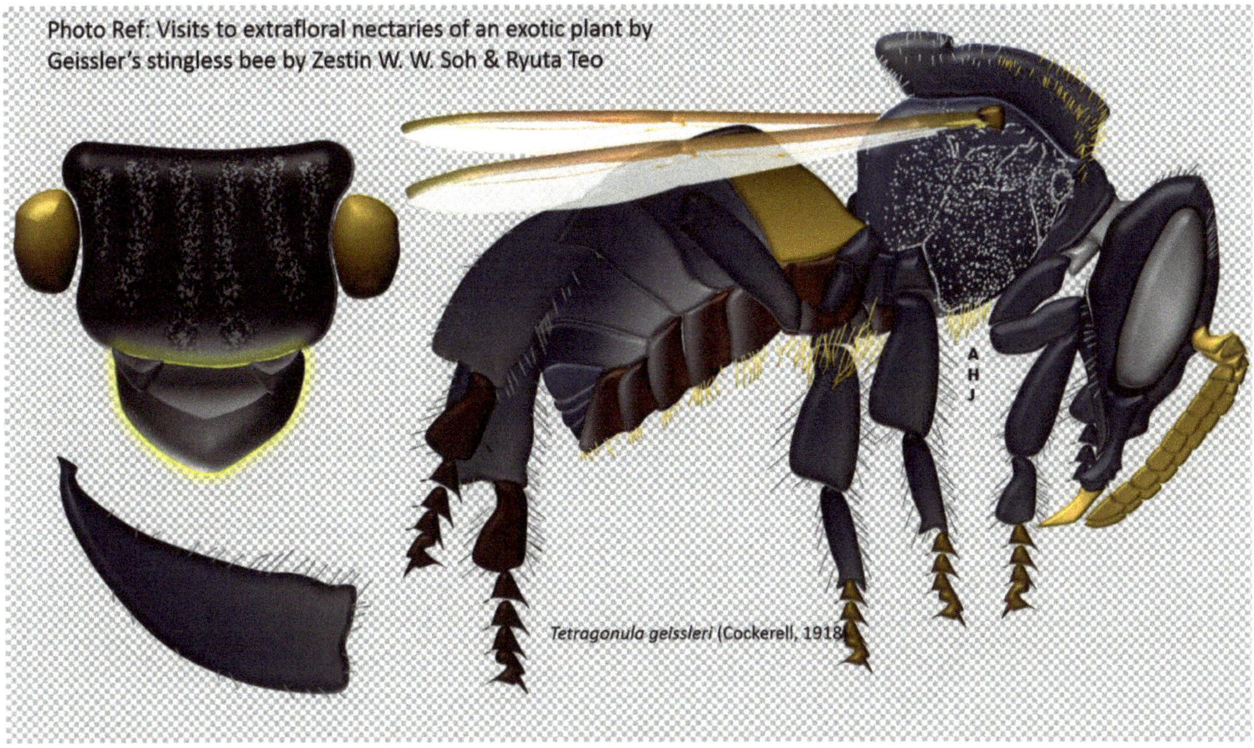

Illustration 12 Tetragonula geissleri (Cockerell 1918)

Cameron Highlands - b) **Melipona basimaculata*[53] Bingham 1903, nomen dubium.

[53] *Tetragonula basimaculata* Generated: Monday, April 5, 2021
https://www.itis.gov/servlet/SingleRpt/SingleRpt?search_topic=TSN&search_value=1129864#null/

Melipona basimucalata Bingham 1903

♂ *Melipona basimaculata* Bingham 1903, nomen dubium et inquirendum
Digital rendered from silhouette images and descriptons in Bingham 1903
Habitat: Biserat, Jalor ; Telom, Perak-Pahang boundary, Malaysia. 4,ooo feet.

*Illustration 13 *Melipona basimaculata Bingham 1903*

Label 6 Melipona basimaculata Bingham 1903

Reference for:	*Tetragonula basimaculata*
Acquired:	2019
Notes:	Ascher, J. S. and J. Pickering. 11 October 2019. Discover Life bee species guide and world checklist (Hymenoptera: Apoidea: Anthophila). http://www.discoverlife.org/mp/20q?guide=Apoidea_species

7.2 Larut Hills - *Tetrigona apicalis* (Smith, 1857)

Label 7 Tetrigona apicalis (Smith, 1857)

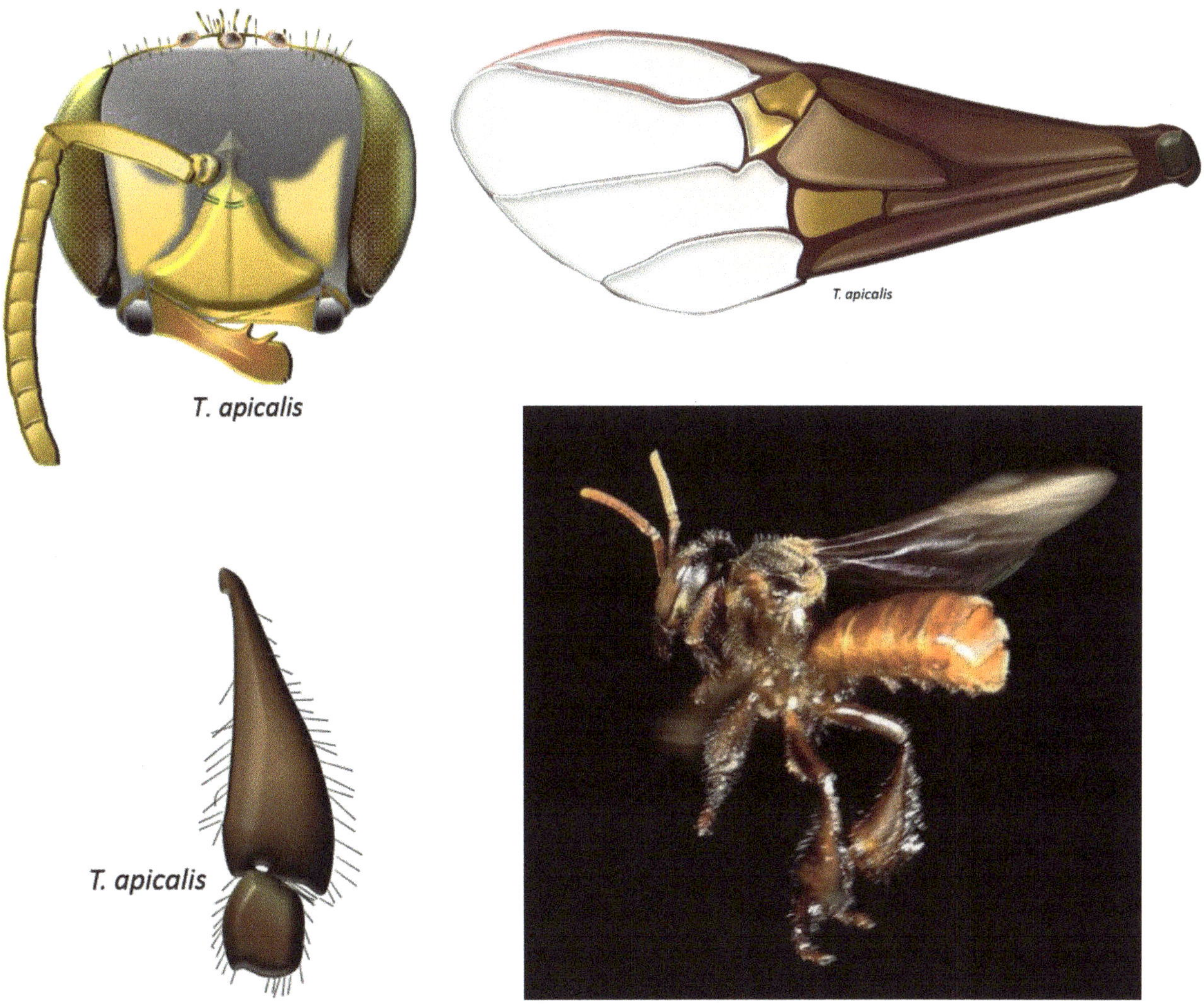

Illustration 14 Tetrigona apicalis (Smith, 1857)

7.3 Mt Dulit -

Label 10 Borneotrigona hobbyi

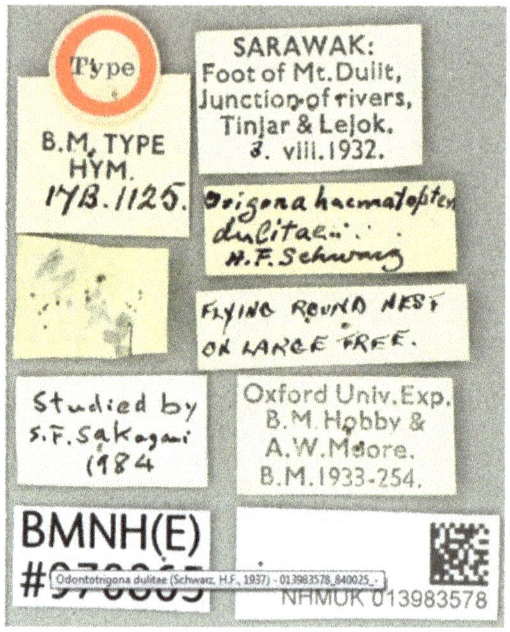

Label 11 Odontotrigona haematoptera

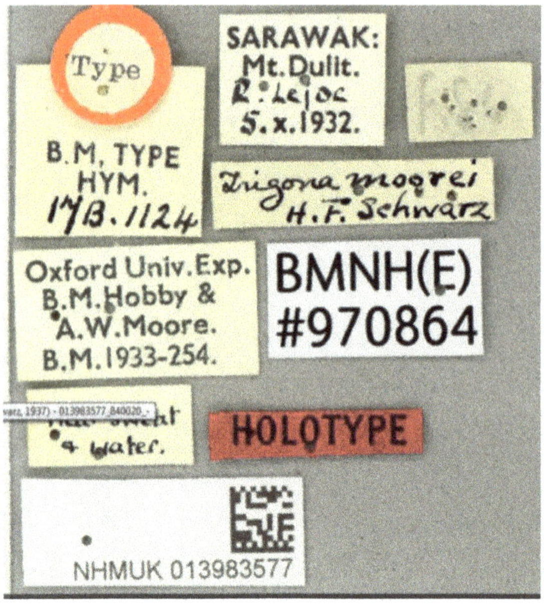

Label 9 Sundatrigona moorei

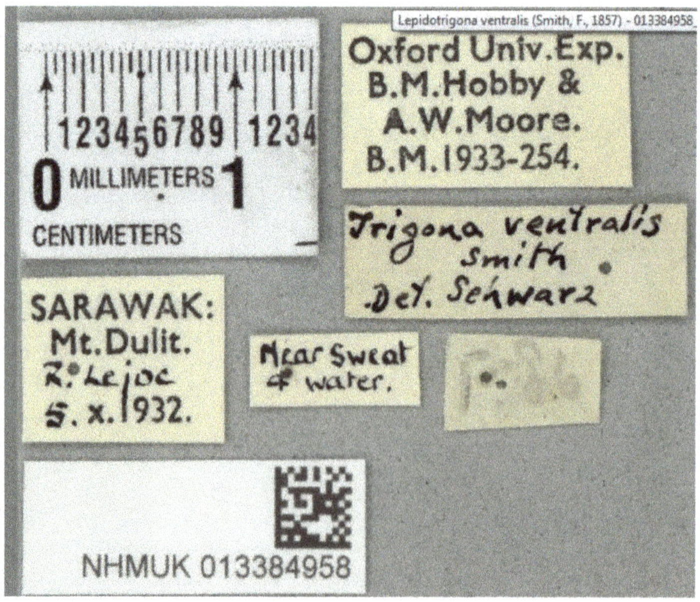

Label 8 Lepidotrigona ventralis

[1937] Schwarz, Oxford Univ. Sarawak (Borneo) Expedition: Trigona p282

List of the species taken at different elevations on Mt. Dulit with new combination synonyms bold in parenthesis.

7.3.1-Lowland Slopes

A.-Foot of Mt. Dulit

Trigona confusella [***Tetragonula geissleri (Cockerell, 1918)***]

Trigona melanocephala [***Tetragonula melanocephala*** **(Gribodo, 1893)**]

Trigona haematoptera variety *dulitae* [***Odontotrigona haematoptera*** **(Cockerell, 1919)**]

Trigona terminata variety *latebalteata*, new variety [***Lepidotrigona latebalteata*** **(Cameron, 1902)**]

Trigona iridipennis [***Tetragonula iridipennis*** **(Smith, 1854)**]
Trigona thoracica variety *lacteifasciata* [***Geniotrigona lacteifasciata*** **(Cameron, 1902)**]
Trigona itama [***Heterotrigona itama*** **(Cockerell, 1918)**]
Trigona ventralis [***Lepidotrigona ventralis*** **(Smith, 1857)**]

B.-Mt. Dulit, R. Lejok
Trigona iridipennis [***Tetragonula iridipennis*** **(Smith, 1854)**]
Trigona melina [***Tetragonula melina*** **(Gribodo, 1893)**]
Trigona melanocephala [***Tetragonula melanocephala*** **(Gribodo, 1893)**]
Trigona moorei, new species [***Sundatrigona moorei*** **(Schwarz, 1937)**]
Trigona ventralis [***Lepidotrigona ventralis*** **(Smith, 1857)**]

C.-Mt. Dulit Trail
Trigona fimbriata [***Homotrigona fimbriata*** **(Smith, 1857)**]
Trigona melanocephala [***Tetragonula melanocephala*** **(Gribodo, 1893)**]
Trigona iridipennis [***Tetragonula iridipennis*** **(Smith, 1854)**]
Trigona melina [***Tetragonula melina*** **(Gribodo, 1893)**]

7.3.2 Mt. Dulit, At Elevation 2500 Feet
Trigona apicalis [***Tetrigona apicalis*** **(Smith, 1857)**]
Trigona collina [***Tetragonilla collina*** **(Smith, 1857)**]
Trigona melina [***Tetragonula melina*** **(Gribodo, 1893)**]

7.3.3 Mt. Dulit, At Elevation 4000 Feet
Moss Forest
Trigona collina [***Tetragonilla collina*** **(Smith, 1857)**]
Trigona melanocephala [***Tetragonula melanocephala*** **(Gribodo, 1893)**]
Trigona hobbyi, new species [***Borneotrigona hobbyi*** **(Schwarz, 1937)**]
Trigona rufibasalis [***Tetragonilla rufibasalis*** **(Cockerell, 1918)**]
Trigona sarawakensis, new species [***Tetragonula sarawakensis*** **(Schwarz, 1937)**]

7.3.4 R. Kapah, Tributary of R. Tinjar
Trigona fimbriata [***Homotrigona fimbriata*** **(Smith, 1857)**]
*Trigona itama [**Heterotrigona itama*** **(Cockerell, 1918)**]
Trigona melanocephala [***Tetragonula melanocephala*** **(Gribodo, 1893)**]

7.3.5 Mt. Kalulong, At Elevation 1800 Feet
Trigona haematoptera variety *dulitae,* [***Odontotrigona haematoptera*** **(Cockerell, 1919)**]

7.4 Batang Padang, Tg Malim - *Tetrigona peninsularis*

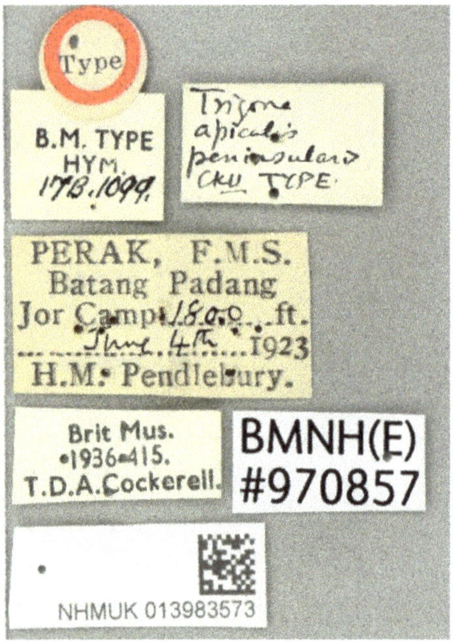

Label 12 Tetrigona peninsularis

7.5 Sungai Pinang Hill - *Tetragonula iridipennis*

Label 13 Tetragonula iridipennis

7.7 Mt Ophir (Gn. Ledang)

Those of us who are familiar with the Scriptures may also perhaps recall Solomon's legendary goldmines of Ophir and the attribution of its location by recent scholars to our Gunung Ledang. Solomon reportedly sent Hiram's fleet to Ophir to obtain incense, sweet-woods and gold for his temple, produce formerly abundantly available from Malayan shores. If nothing else, this quaint hypothesis at the very least acknowledges a passing familiarity in the Levant with our region in Biblical times. (Raimy Ché-Ross, Barton, Canberra 2002)

One specimen in The Natural History Museum, London (''BM 17B 1184'') is labelled as and was long considered as the type specimen of *T. laeviceps*. However, it is not from Singapore but is labelled Mt. Ophir (= Gunung Ledang), an isolated mountain in Malaysia nearly 200 km northwest of Singapore. Because of the locality, this specimen cannot be the holotype or a syntype of *T. laeviceps*. It is extremely dirty so that it is impossible to see many surfaces and setal features. (Rasmussen & Michener 2010)

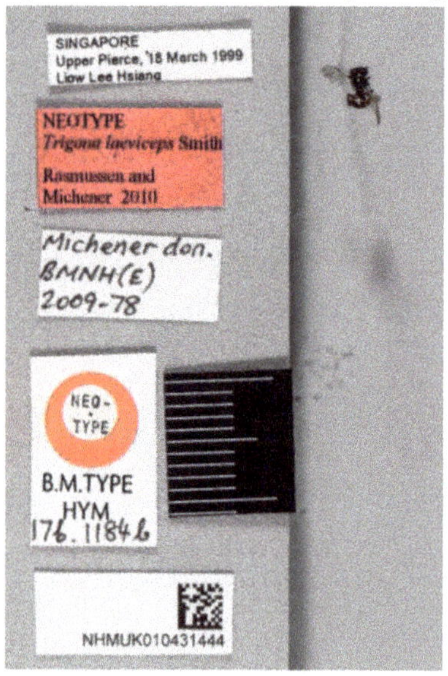

Label 14 Tetragonula laeviceps

7.7.1 Expedition

In the British Museum is a specimen from the Smith collection obtained at Mt. Ophir, the type locality of nitidiventris. It, too, is designated nitidiventris and has the word "type" written on the accession label... *nitidiventris* Smith, from Mt. Ophir, which I have seen in the Saunders collection at Oxford. This specimen of *nitidiventris* is labelled "M. Ophir 79" (Excerpt from Schwarz 1937 pg. 292)

With these notes, Dr. Hans Bänziger contacted me and expressed his desire to visit Mt. Ophir to search for *Lepidotrigona nitidiventris.* I relayed his intentions to friends in The Malaysia Genome Institute to attain formal clearance with the Gunung Ledang Park management for a scientific visit. Arrangements set, and we organized an expedition with some colleagues and enthusiasts and headed off to the Gn Ledang Park in November of 2015. It was a 2-hour drive, and we set base at a nearby motel in Tangkak.

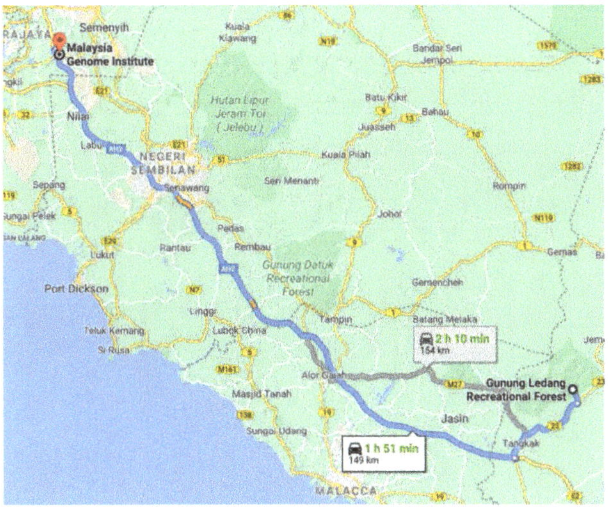

Nap 1 Route from MGI to Gn. Ledang

Figure 45 Ledang expedition team

Members of the expedition: seated 1st row from left, Mohd Noor (Malaysia Genome Inst) Dr, Hans Bänziger (Chiangmai University), Abu Hassan Jalil (Akademi Kelulut Malaysia), Mohd Zaki (AKM), Mohd Razif Mamat (MGI), ; 2nd Row from left, Ms. Anna (UKM), Sangdao Bänziger Halimah (MGI) , Intan & Jeffry Iskandar Foo (AKM)

Figure 47 Tetragonilla atripes entrance tube (Photo by Jeffry) ; members of the expedition: seated 1st row from left, Mohd Noor (Malaysia Genome Inst) Dr, Hans Bänziger (Chiang Mai University), Abu Hassan Jalil (Akademi Kelulut Malaysia), Mohd Zaki (AKM), Mohd Razif Mamat (MGI), ; 2nd Row from left, Ms. Anna (UKM), Sangdao Bänziger Halimah (MGI) , Intan & Jeffry Iskandar Foo (AKM)

Figure 46 Top left: Razif examining specimen; Top right and bottom left: entrance tubes discovered; Bottom right: Dr. Bänziger checking out the entrance tube

Figure 48 Team hard at work

From detailed discussions with Dr. Hans Bänziger (via Sangdao CMU) after the expedition organized by the Academy of Meliponiculture and Malaysian Genome Inst, there appears to be inconsistency. Identification based on misinterpretations of the key descriptions compared to designations practised at Chiang Mai University may differ. Possibly, the interpretations by AKM and MGI may not be in line with the names that are applied by taxonomists or in the collections at the university there. The problematic interpretations are in the Genus *Lisotrigona*, *Tetragonula* and *Tetrigona*.

The recognition methods in Malaysia / Brunei can be easily reviewed[54]. AKM works on updates on composite key algorithms, and MGI is working on DNA sequencing and analysis.

[54] Editor's note: If there are good specimens and somebody with a microscope, there is little reason for doubt.

Figure 49 Relentless and determined.

Figure 50 Found something and yet to be detailed

7.8 Fraser's Hill

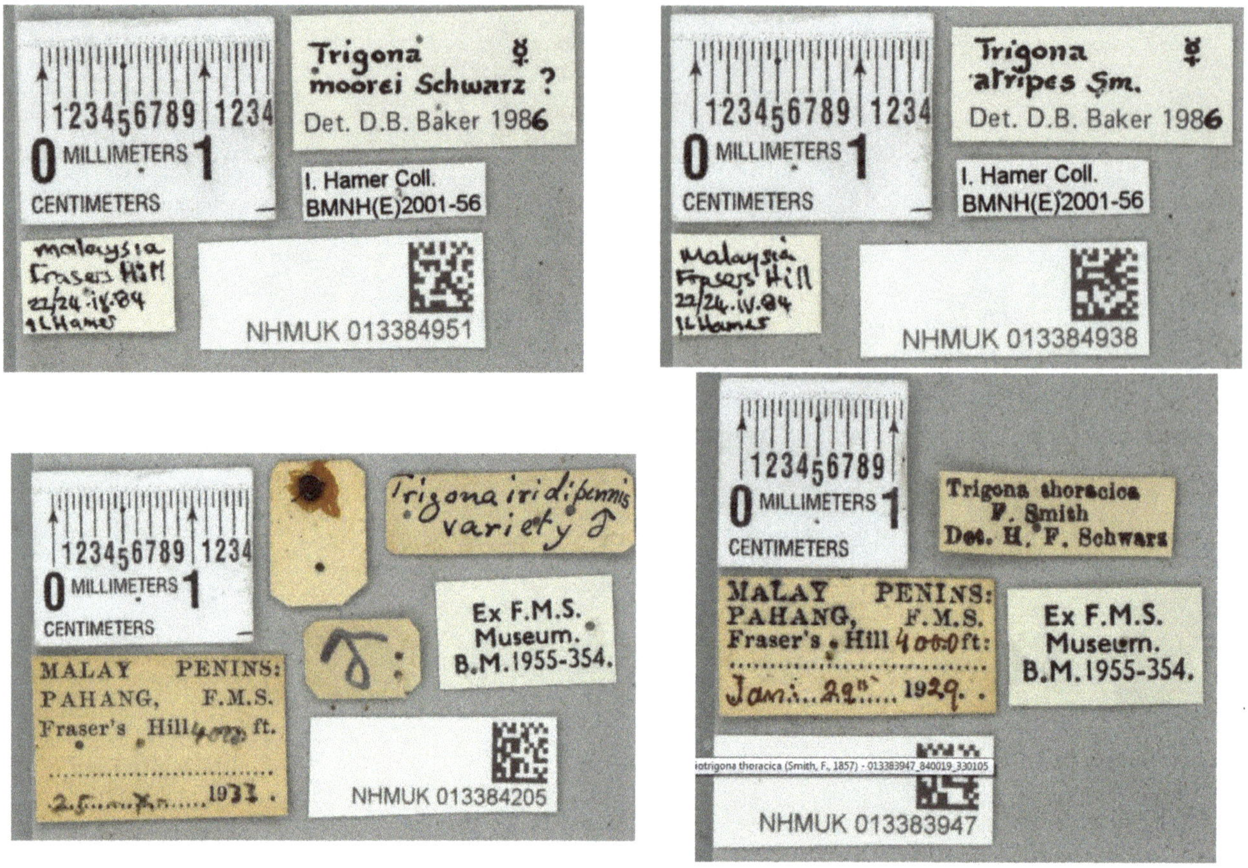

Figure 51 Collective Labels of S. moorei; T. iridipennis; T. atripes & Geniotrigona thoracica

7.9 Kancing Hill, Selangor

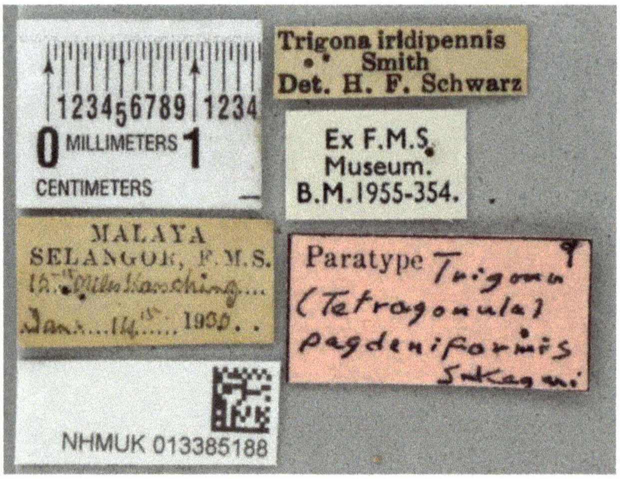

Label 15 Tetragonula pagdeniformis

7.10 Bukit Ketam, Tuaran, Sabah

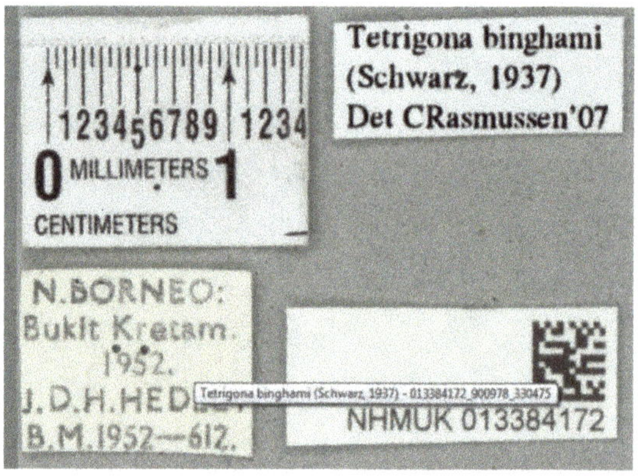

Label 16 Tetrigona binghami

7.11 Kedah Peak (Gn Jerai)

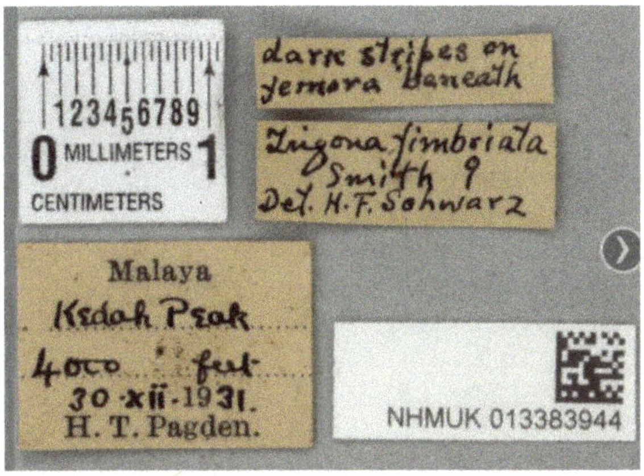

Label 17 Homotrigona fimbriata

7.12 Bukit Kutu, Selangor

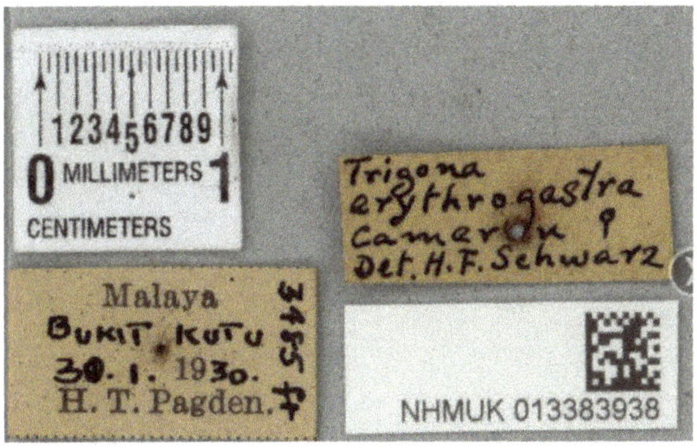

Label 18 Bukit Kutu - Heterotrigona erythrogastra

7.13 Ulu Gombak- Forest Reserve

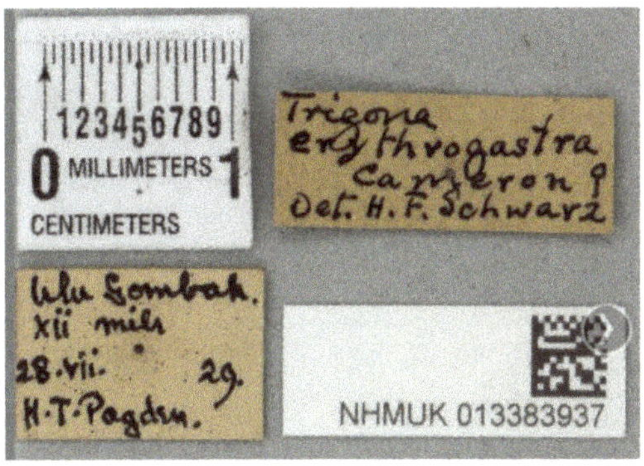

Label 19 Ulu Gombak - Heterotrigona erythrogastra

7.14 References

The Indo-Malayan Species of Stingless Bees by Herbert F. Schwarz. Bulletin of The American Museum of Natural History, Vol. LXXVI, Art III, New York, issued July 31, 1939.

Results Of The Oxford University Sarawak (Borneo) Expedition: Bornean Stingless Bees Of The Genus *Trigona* by Herbert F. Schwarz, Bulletin of The American Museum of Natural History, Vol. LXXIII, Art III New York, issued May 3rd, 1937

An open-source project by the Natural History Museum's Biodiversity Informatics Group. Data Portal https://data.nhm.ac.uk/

List of NHUK Museum specimen labels.

Label 1 Tetragonula pagdeniformis (Sakagami, 1978) 117
Label 2 Tetragonula iridipennis (Smith, 1854) 120
Label 3 Tetragonula melina (Gribodo, 1893) 121
Label 4 Lepidotrigona terminata (Smith, F., 1878) 124
Label 5 Tetragonula geissleri (Cockerell 1918) 137
Label 6 Melipona basimaculata Bingham 1903 138
Label 7 Tetrigona apicalis (Smith, 1857) 139
Label 8 Lepidotrigona ventralis 140
Label 9 Sundatrigona moorei 140
Label 10 Borneotrigona hobbyi 140
Label 11 Odontotrigona haematoptera 140
Label 12 Tetrigona peninsularis 142
Label 13 Tetragonula iridipennis 142
Label 14 Tetragonula laeviceps 142
Label 15 Tetragonula pagdeniformis 148
Label 16 Tetrigona binghami 148
Label 17 Homotrigona fimbriata 149
Label 18 Bukit Kutu - Heterotrigona erythrogastra 149
Label 19 Ulu Gombak - Heterotrigona erythrogastra 149

List of Illustrations

Illustration 1 Tetragonula penangensis (Cockerell, 1919) 112
Illustration 2 Tetragonilla atripes (Smith, 1857) 113
Illustration 3 Tetragonula fuscobalteata (Cameron, 1908) 115
Illustration 4 Tetragonula laeviceps (Smith, 1857) 116
Illustration 5 Male Tetragonula pagdeniformis (Sakagami, 1978) 117
Illustration 6 Tetragonula zucchii (Sakagami, 1978) 118
Illustration 7 Tetragonula iridipennis (Smith, 1854) 120
Illustration 8 Tetragonula melina (Gribodo, 1893) 122
Illustration 9 Tetragonilla collina (Smith, 1857) 123
Illustration 10 Lepidotrigona terminata (Smith, F., 1878) 124
Illustration 11 Heterotrigona bakeri (Cockerell, 1919) 124
Illustration 12 Tetragonula geissleri (Cockerell 1918) 137
*Illustration 13 *Melipona basimaculata Bingham 1903* 138
Illustration 14 Tetrigona apicalis (Smith, 1857) 139

Chapter 8

8.0 Meliponiculture potential in Ranau, Sabah

8.0.1 Introduction

Ranau is noted for its hilly geographical structure and is the largest producer of highland vegetables in Sabah. Tourism and highland agriculture are the major industries, as the district is at 1,176 m above sea level.

Kundasang is a town in the district of Ranau in Sabah, Malaysia, that lies along the bank of Kundasang Valley. It is located about 6 kilometres away from Kinabalu National Park, 12 kilometres from Ranau town and is renowned for its vegetable market, which is open seven days a week. It is the closest town to Mount Kinabalu and has a panoramic view of the Mountain. It is populated mainly by the native Dusun and a small population of Chinese people. Locals operate all the shops.

It is the highest settlement in Malaysia at an elevation of almost 1,900m (6,200ft).
With this significant trivia, it is imperative to review an account of stingless bee diversity in this region and the potential of stingless beekeeping in the region.

John Wyne James is an avid macro photographer living on Ranau. I met him when visiting Kundasang in 2016 as part of a Borneo Meliponiculture tour by the Akademi Kelulut Malaysia group. We were travelling from Tambunan, heading towards Kundasang. We stopped at the Ranai Town food court and arranged to meet John. It was a pleasant meet where we discussed the prospects of publishing a coffee table book. These are the photos he contributed for publication. We also inquired if there were beekeepers in the region and none known of it at a commercial scale. Some hobbyists may exist as bamboo hives sold at the roadside vegetable mart at Kundasng. Following is a Gallery of photographs captured by him:

8.0.2 Photo Gallery

Figure 52 Stingless bees of Ranau, Sabah

Johnwyne James
Freelance
Photography 2015

Johnwyne James
Freelance
Photography 2015

Johnwyne James
Freelance
Photography

Johnwyne James
Freelance
Photography 2015

8.1 Diversity of Stingless bee Species in Ranau, Sabah.

by Rusphin Asyraf Taip

Ranau is a district located in the North West Coast Division of Sabah, Malaysia. It is estimated that the total population in 2000 was 70,649 people. Ranau district is located between Latitude 5 ° 30' N and 6 ° 25' N and longitude 116 ° 30 ' E and 117 ° 5' E. The area of Ranau district is 294,224 hectares or 1978.5 km square. Three main road links connect it with cities and districts such as Kota Kinabalu, Sabah East Coast and Inland. The roads are Ranau-Tamparuli Road, Ranau-Telupid Road and Ranau-Tambunan Road. It is a well-known tourist destination in the country and internationally. Among the tourist attractions are Poring Hot Spring, Mount Kinabalu National Park, Sabah Tea Farm, Kundasang Town (which is famous for its vegetable stalls), former Mamut copper mine, World War II prison (historical prisoner-of-war camp), Sungai Liwagu (a destination for water rafting activities), monthly "Tamu" market (once a month) and many more other attractions.

Ranau is a hilly district located in the Crocker Range and the Pinousuk Peak in the North, the Ranau Valley in the East and the Trus Madi Range, and the Labuk Highlands in the South. The main drainage is Sungai Liwagu and Sungai Sugud. In short, Ranau District is rugged with valley plains, undulating land and mountainous terrain in most areas. Ranau has a humid yet relatively cool climate. Ranau is a breeding ground for several stingless bee species with its hilly terrain and rich in natural forest resources. Throughout the study from 2016-2020, there are species of Indo-Malaya stingless bee that live here, such as *Geniotrigona, Heterotrigona, Lepidotrigona, Lophotrigona, Sundatrigona, Tetragonilla*, *Tetrigona* and *Tetragonula,* and to review the potential in Highland Meliponiculture.

8.1.1 Genus *Geniotrigona*

Throughout the research conducted in Ranau, the species *Geniotrigona lacteifasciata* were found in several places along the Trus Madi Range and the Sugud river in the Labuk Highlands.

Fig 1 Geniotrigona lacteifasciata

Fig 2 Conspicuous orange coloured clypeus.

8.1.2 Genus *Heterotrigona*

There are two species in this genus that can be found in this area. *Heterotrigona itama* is found in many locations. This species is the choice of many stingless bee breeders because it produces a lot of honey. This species is found nesting in the cavities of tree trunks and bamboo. The second species is *Heterotrigona erythrogastra.* Throughout the research conducted, *H. erythrogastra* only had its presence found only in the forest area of the Labuk Highlands along the hot Sugud river. To date, there is no record of it found in the cooler regions in the Crocker Range.

Fig 3 H. itama

Fig 4 H. erythrogastra

8.1.3 Genus *Lepidotrigona*

Throughout four years of research in this area, several species identified of this genus were found. First and foremost is the species *L.* cf.[55] *latebalteata* and *L.* cf. *javanica, L.* cf. *ventralis* and *L.* cf. *doipaensis.* All of these species were easily found in this area. Still, observations indicate *L.* cf. *latebalteata* and *L.* cf. *ventralis* is found abundantly in the Crocker Range's upland areas. Along with the river Liwagu and the river Sugud, it is relatively easy to find species *L.* cf. *javanica* and sometimes *L. ventralis* in this area. *L. doipaensis* was discovered in the area near Kundasang, and Mr. Abu Hassan Jalil identified it from close-up photos.

Fig 6 Lepidotrigona cf. latebalteata

Fig 5 Lepidotrigona cf. ventralis

Fig 7 Lepidotrigona cf. javanica

Fig 8 Lepidotrigona cf. doipaensis

[55] Editor's note: the expression 'cf.' (it takes a period) means that the identification is tentative- something seemed different from the species familiar to the author

8.1.4 Genus *Tetrigona*

Dipterocarp forests along the Liwagu River and the Sugud river give life to many stingless bees that depend on the resin. The most abundant species found in this area is *T. binghami*. Observations done for four years show that this species is found in abundance along the area bordering the district of Kota Marudu and the district of Beluran. While the species *T. melanoleuca* was also identified in this location, but its existence is not much like the species *T. binghami.*[56] This species is easily found in this location because most of the area still retains Dipterocarp forest.

Fig 9 Tetrigona binghami

Fig 10 Tetrigona melanoleuca

8.1.5 Genus *Tetragonilla*

This observation shows *T. collina*[57] species found in all dipterocarp forest areas either in the Trus Madi Range and the Crocker Range, and the Labuk Highlands. It is possible that there are other species in this genus but not yet found. This species was also reported in the Lohan area and adapted to surrounding sources without Dipterocarp trees. The bee was also found at Poring at the foot of Mount Kinabalu.

Editor's notes:

[56] I was pretty sure *binghami* from Brunei was black, and its entrance tube was long and narrow...

[57] The main things to say about *T. collina* are that its colonies are made underground at the base of large trees and often occur there in groups of several colonies.

Fig 12 Tetragonilla collina

Fig 11 Nest structure of T. collina

8.1.6 Genus *Lophotrigona*

Before this observation, most *L. canifrons*[58] were found in forest areas along the Sugud river, especially in Malinsau, near Kota Marudu District's border.

Fig 13 Lophotrigona canifrons

Fig 14 Nest structure of L. canifrons

[58] Editor's notes: Add that it is without doubt one of the most highly defensive stingless bees in Asia, as all those who molest a nest have discovered.

8.1.7 Genus *Sundatrigona*

In 2018, an expedition conducted by Prof Shamsul Bahri Abd Razak to track *S. moorei* took place in Penampang, Tamparuli, Keningau and Ranau. This expedition's results revealed to the outside world that *S. moorei* is found in abundance there. This stingless bee nests in the nest of *Crematogaster* ants. In Ranau, this stingless bee was found at an altitude of 1000 feet above sea level, especially in the hilly areas and cooler temperatures along with the Crocker range. Throughout the research conducted in this area, this species was not found in areas along the Liwagu River and Sugud River.

Fig 18 Crematogaster sp.

Fig 17 S. moorei in an ant nest

Fig 15 Nest entrance of S. moorei

Fig 16 brood cell structure of S. moorei

8.1.8 *Borneotrigona hobbyi*

The species in this genus were seen for the first time in 2018. After research conducted with Mr. Abu Hasan Jalil, we found that this species is most likely aff. *Borneotrigona hobbyi.*

**Note:* The nest entrance aperture shape of this species and *binghami* or *fuscibasis* is in medical terminology. "Introitus" means in the form of the vagina or vaginal opening. Thus, speculatively it is associated with traditional female medicine, for *apicalis.* It is in the form of a 'keris' (dagger) scabbard. For *thoracica,* it is in the form of the anus, so it is associated with haemorrhoid medicine. This has nothing to do with the actual practice of the traditional remedy. It is just a speculation of the medicinal value of the propolis material of the species. There is no authentic evidence.

*Fig 19 *nest entrance of B. hobbyi*

Fig 20 B. hobbyi worker

8.1.9 Genus *Tetragonula*

Tetragonula melina was seen in Malinsau in 2017 and in Kemburongoh near Kundasang in 2019, foraging in cool uplands. *T. melanocephala* were also found in abundance in these areas. This bee's distribution followed its existence either in the cool upland areas or areas along the Liwagu River and the Sugud River in warmer temperatures. Apart from that, *T. pagdeni* is also found in the Malinsau area in Sungai Sugud and reported in several other locations such as Lohan, around Ranau town and the cool temperature area of Kemburongoh near Kundasang.

The species *T. fuscobalteata*[59] accepts a wide variety of habitats, and they are found in abundance throughout Ranau. This species spreads from cold upland areas to hot temperatures.

Tetragonula reepeni and *T. laeviceps* are found easily around the Ranau area. These species are found in the highlands in cool areas as well as in warmer spots.

In 2019, Prof. Shamsul Bahri Abd. Razak and his entourage from UMT visited Ranau to find *T.* cf. *sarawakensis* in Sumalang. He managed to take macro photos and get some samples of this species as study material. Some more species in the genus *Tetragonula* were found there. However, due to limited information and research, the species were unidentified. Similarly, there are several other species in other genera that have not been identified but recorded in the Ranau district. Among them are

Fig 26: T. pagdeni

Fig 25: T. cf. reepeni

Fig 24 : T. laeviceps

Fig 23 T. melina

Fig 22 T. melanocephala

Fig 21 T. fuscobalteata

Tetragonilla rufibasalis, Odontotrigona haematoptera and others. We hope that other researchers can verify the species ID of stingless bees in this area. Many thanks to Mr. Abu Hassan Jalil for his help in writing this article, especially for species identification.

[59] *Tetragonula fuscobalteata* is one of the smallest Asian meliponines (excluding the tiny Meliponini- *Lisotrigona* and *Pariotrigona*).

8.1.10 *Odontotrigona haematoptera*

Species are found in Ranau and are rather rare in that area.

Fig 27 a: lateral habitus, b: dorsal habitus (The red wings and angular shape of hind tibia implies it could be aff O. haematoptera, predicted by Abu Hassan Jalil)

Fig 27 a-b: This bee is not found in its nest to date. It is seen several times coming to collect pollen during the log cutting process done in Malinsau. (The red wings and angular shape of hind tibia implies it could be aff. *O. haematoptera,* as suggested by AHJ*)*

8.1.11 Isolated Specimens

Some worker specimens were found foraging in the wild. Without their nests, it was difficult to determine the type.

Fig 28 a species of the genus Lepidotrigona because it has yellow scales on its scutellum. (suspected L. aff. latipes by Abu Hassan Jalil)

Fig 28 a-b: This bee was discovered foraging in 2020 in Malinsau. The nest was not found. Close inspection of the specimen shows that it is most likely a species of the genus *Lepidotrigona* because it has yellow scales on its scutellum. (suspected *L.* aff. *latipes* because of its extremely broad hind tibia)

Fig 29: Another specimen found and isolated. AHJ suspects this as aff. *Tetragonula rufibasalis*. This bee is scarce, and the nest is yet to be found.

Fig 29 worker specimen. (possibly aff. T. rufibasalis as suggested by AHJ)

8.1.12 Unidentified Types

The following nest entrances have not yet been identified.

Fig 30 Selection of nest entrances unidentified yet.

List of Fig in chapter 8

Fig 1 Geniotrigona lacteifasciata 155
Fig 2 Conspicuous orange coloured clypeus. 155
Fig 3 H. itama 155
Fig 4 H. erythrogastra 155
Fig 5 Lepidotrigona cf. ventralis 156
Fig 6 Lepidotrigona cf. latebalteata 156
Fig 7 Lepidotrigona cf. javanica 156
Fig 8 Lepidotrigona cf. doipaensis 156
Fig 9 Tetrigona binghami 157
Fig 10 Tetrigona melanoleuca 157
Fig 11 Nest structure of T. collina 158
Fig 12 Tetragonilla collina 158
Fig 13 Lophotrigona canifrons 158
Fig 14 Nest structure of L. canifrons 158
Fig 15 Nest entrance of S. moorei 159
Fig 16 brood cell structure of S. moorei 159
Fig 17 S. moorei in an ant nest 159
Fig 18 Crematogaster sp. 159
*Fig 19 *nest entrance of B. hobbyi* 160
Fig 20 B. hobbyi worker 160
Fig 21 T. fuscobalteata 161
Fig 22 T. melanocephala 161
Fig 23 T. melina 161
Fig 24 : T. laeviceps 161
Fig 25: T. cf. reepeni 161
Fig 26: T. pagdeni 161
Fig 27 a: lateral habitus, b: dorsal habitus (The red wings and angular shape of hind tibia implies it could be aff O. haematoptera, predicted by Abu Hassan Jalil) 162
Fig 28 a species of the genus Lepidotrigona because it has yellow scales on its scutellum. (suspected L. aff. latipes by Abu Hassan Jalil) 162
Fig 29 worker specimen. (possibly aff. T. rufibasalis as suggested by AHJ) 163
Fig 30 Selection of nest entrances unidentified yet. 164

Chapter 9

9.0 Merlimau Kelulut Honey Farm

by Muhammad Fahdli Abdul Rahman

9.1 Introduction

The integration between kelulut and coconut trees in Sungai Rengi 2 Village, Merlimau, Melaka, was inspired by Mr. Muhammad Fahdli Abdul Rahman's idea, 28 years old. He is a graduate majoring in Biochemistry at the bachelor's level in the analysis of Kelulut honey and is doing his PhD in plant science.

9.2 Establishment

Figure 53 Merlimau Kelulut Honey Farm

This area is located approximately 6.3 km from Merlimau (10-minute drive), and about 25 kilometres of the Malacca state airport is a 30-minute drive. In Merlimau Kelulut Honey Farm, there are two types of kelulut nests: logs and hives (boxes). They are also used as garden landscape features because each of the stingless bee nests is set up on carved floral motif stone pedestals.

The owners' concept of meliponiculture is keeping the main nest in the cut tree trunk stumps or logs placed on pedestals for the appropriate, workable height.

Figure 54 Qualified staff

The stingless bee type cultured is also the most popular kelulut, i.e. *Heterotrigona. itama* and are very suitable for those who want to start kelulut farming. Then wooden boxes, also known as 'supers' or, as in this case ', topping' hive boxes, are placed connected on top of the nest logs. These topping boxes serve as honey storage pot compartments are created using lumber and planks.

Above these topping boxes, we may put suitable roofing to shelter the whole hive setup from the rain and direct sun. All these setups are then placed, collectively in clusters of dozens or so, under shady trees or netted roof structures to maintain suitable ambient temperatures and humidity.

To choose a robust and hardy type of stingless bee for meliponiculture is by inspection of the entrance funnel or tube. Stingless bee nests have an entrance funnel built for the bees going in and out of the nest. The materials used for constructing the entrance funnel include propolis, wax, sand, resin, soil or sand particles and plant exudates. The shape and the texture of these entrance funnels permit the recognition of stingless bee species.

The Merlimau Kelulut Honey Field has 1.5 acres with 100 colonies and nearly 140 coconut trees. Coconut flowers are the primary source of food for the bees here, and the owners provide these coconut trees as 60-70% of the nectar forage source for honey production. Several other trees complement the development of stingless bee honey in this field as "minor food supply", such as mango, mangosteen, 'chempedak' and jackfruit, guava and flowering trees, including Arabica coffee, longan and durian trees. Some flowers have high nectar, including flowers, 'bridal tears' (*Antigonon leptopus*), Gourds (Cucurbits), Basil, Jambos (*Syzygium*), and also Jasmine.

Figure 55 Harvesting Honey in full attire.

9.3 Production

Figure 56 Meliponiculture in coconut plantation

Stingless bee honey produced in the Kelulut Merlimau field can be said to be monofloral (coconut) honey. The taste and aroma of honey are consistent compared to other areas, which are more multi-floral. Monofloral honey from the Merlimau kelulut field is mostly coconut nectar. However, the methods of honey suction and controlling the hive constitute a standard operating procedure. The cleanliness practised and the correct and appropriate harvesting methods preserve the quality of the honey. The honey harvest records were taken every 30-40 days, while the hives were inspected once a week.

However, unmaintained but constantly producing farms will be a natural attraction for wild animals, or other enemies and predators to *H. itama* colonies, to invade for survival. The hive or honey and propolis are vulnerable to pillage, predation or destruction by the enemies like predatory arthropods, ants, termites, and other animals such as lizards, frogs, monkeys, bears, wild boars and so on.

Besides, Malaysia is a tropical country that faces only two seasons of hot and rainy only (Monsoon & Inter-Monsoon). The impression of climate change and bad weather conditions dramatically affects the development and habitat of kelulut. For example, when the weather is too wet and humid, the rain continues to make stingless bee colonies less active. The hot and dry weather conditions give the stingless bee colony the challenges of increased competition from other foragers and dehydration due to a lack of water resources. Due to weather changes and not well-controlled conditions, it will lead to deterioration and loss of colonies. The main factors for ensuring colony resilience are a variety of food and resin sources and a consistent water provision.

Figure 57 Visit Melaka Year 2019 – VIP formal visit

9.4 Current Status

So far, honey production at Kelulut Merlimau Honey Field always continues. The profit from selling kelulut honey and kelulut nests is also very stimulating. The average new colony duplication from splitting is 2 - 3 nests in one month.

To everyone's knowledge, the Merlimau Kelulut Honey Field was selected by Tourism Melaka to be one of the promotional places for Visit Melaka Year 2019. Visitors from outside and inside the country can observe the Meliponiculture and maintenance of kelulut on this farm.

The Chief Minister of Melaka, YAB Tuan Hj. Adly bin Zahari ('DUN' Merlimau – State Legislative Assemblyman) on September 26, 2019, visited the stingless bee field. During this visit, he was intrigued about the integration of kelulut in coconut fields, the increased yields of the coconuts and improved quality of the fruit and how kelulut helps in coconut pollination.

Figure 58 Participants of Advanced Meliponiculture course

Merlimau Kelulut Honey Farm is sometimes used for discourse and discussions on matters of kelulut within and outside the country and to increase knowledge and increase the potential for Meliponiculture in Malaysia.

9.5 Aspirations

Throughout 2019, the Merlimau Kelulut Honey Farm has received visits and visits from kelulut clubs and associations, state agricultural departments, the officials of the chief minister department of Melaka, and staff from Melaka Halal Hub Sdn. Bhd., as well as university students and the general public from all over the country. They liked the idea of implementing integrated farming in the place. The next mission is to develop and expand Merlimau Kelulut Honey Farm. Future projects will help open up employment opportunities and, in a way, improve the economy of the agricultural sector in the country. We aim to explore new directions in the kelulut honey industry.

Figure 59 PIONEERING INTEGRATED BIOECONOMICS Tags

Chapter 10

10.0 Development of Meliponine Repositories in The Fraser Hill UKM Research Centre And 'Kelulut' Species Identification through DNA Barcoding and Comparison of Honeys

Nor Azfa Johari1*, Mohd Razif Mamat1, Azira Muhammad1, Mohd. Afiq Aizat Jauhari2, Halimah Alias1, Mohd Noor Mat Isa1, Norela Sulaiman2

1Malaysia Genome Institute, National Institutes of Biotechnology Malaysia, Ministry Science, Technology and Inovation (MOSTI), Jalan Bangi, 43000 Kajang, Selangor

2Department of Biological Sciences and Biotechnology; Fraser`s Hill Research Centre, Faculty Science and Technology, 43600 Universiti Kebangsaan Malaysia, Bangi, Selangor

**Correspondence:* azfajo@nibm.my

10.1 Introduction

The Bukit Fraser UKM Research Centre (PPBF) was established in 1990, following the handover of a plot of land from the Kingdom of Pahang in 1989 with an area of 45 hectares covering riparian edges, village heads, rivers and rivers. The position of PPBF at an altitude of 1,002 meters from sea level makes this area a unique flora and fauna repository with a simple ambient temperature range, namely 18 ° C to 22 ° C and high humidity between 85% to 95%. PPBF is also located adjacent to the general recreation area of Jeriau Waterfall, making this location a centre for the strategic investigation of biodiversity still preserved from pollution.

The location of PPBF, which is surrounded by preserved tropical forests, is believed to be the home of various endemic species such as birds, bats and bees. In this study, bees from the Meliponini tribe, or known in Malaysia as Kelulut, were subjected to a review of sustainability and adaptation from lowland to highland areas. Kelulut was reported to be the main common visitor[60] flowering of the staple lowland log species in Malaysia (Momose et al. 1998). Among the characteristics that affect the function of Meliponines are easy compliance and uniformity of flower sources[61] (Ramalho et al., food stores in the nest. They share pollen sources in the nest, reducing the need for workers to come out in 1994). They are maintained and controlled for breeding[62] and harvesting honey. They have large food

Editor's notes:

[60] they are often the common visitors of flowering trees, but they are often not pollinating those trees.

Editor's notes:

[61] They use 100s of flower sources, depending on their location. The average stingless bee or honeybee, in native vegetation, uses around 50 plant species. See Roubik and Moreno, 2013, in the Pot-Honey book.

Editor's notes

stores in the nest. They share pollen sources in the nest, reducing the need for workers to come out in the rainy season[63]. The formation of forager troops through transferring source information to maintenance workers in the nest allows the collection of nectar and pollen from the same source for forage by other workers from the same nest[64] (Roubik 1989).

In Malaysia, however, less information has been obtained regarding Meliponine ability to colonize highland regions. Salim et al. (2012) have reported *Sundatrigona moorei* Schwarz's discovery in the Ulu Gombak Forest Reserve at an altitude of 457 to 1128 meters above sea level. Initial observations have also encountered an unknown Meliponine species on Bukit Fraser at an altitude level of 1,220 meters capable of self-sufficiency in a range of lower than usual temperatures, which is between 17 ° C to 25 ° C. Following these findings, this study was designed to build a PPBF Meliponine repository, identify the surrounding Meliponine species through molecular techniques (DNA barcoding)[65] and compare the metabolite profiles of Meliponine honey from lowland areas and after adapting to highland regions.

10.2 Materials and Procedures

10.2.1 Development of the Meliponine Bukit Fraser Investigation Centre Repository

10.2.1.1 Provision of colonies were obtained from the source.

Meliponine colonies were obtained from sources around Raub and Temerloh, Pahang. Six main species, namely *Heterotrigona itama*, *Geniotrigona thoracica*, *Tetrigona apicalis*, *Tetrigona binghami*, *Lepidotrigona terminata* and *Tetragonula fuscobalteata,* have selected as study species. Topping boxes were installed on top of the log hives and stabilized for 3 to 6 months before the whole setup was moved up to PPBF.

:[62] No real studies on breeding are being done any more. Large collections of bee colonies are deposited in some University campuses and then slowly die. They are artificial populations, have no breeding material, are gradually inbred, and disappear. This has been the fate of virtually all Meliponaries except those in Yucatan in Mexico, where there are local populations of *Melipona beecheii* in the wild (for now) ...

[63] You can say exactly the same thing about honey- except the latter is far more efficient, because the water has been eliminated, and the former has been engineered, through natural selection, to be sustained by the niche-bee coevolution which is continuing now and has been the standard practice for the past 100 million years.

[64] The theme of recruitment in highly eusocial bee colonies has been noted and explained many, many times since the 1950s. It was not just invented by studies in Australia. Try citing an actual compilation or book on the subject- perhaps starting with Lindauer and Kerr, von Frisch, Michener or Seeley and yes, even Roubik (1989). The most recent treatment is by Grüter in a book just published by Springer.

[65] By incorporating molecular data, you have taken a step forward and incorporated another piece in the puzzle. You have not found out what species you are looking at! Barcoding is full of pitfalls and mistakes, due to several different reasons. When it works, then it is just fine. Often there are significant challenges along the way.

10.2.1.2 Provision of site repositories and monitoring of Meliponines

The basis for colony placement in PPBF was built using iron and cement for a sturdy structure and intrinsic barrier. The area around PPBF has also been structured with a bee-friendly landscape by adding floral sources to attract the Meliponines. During the same period, environmental monitoring was carried out to show if Meliponines were presently visiting the garden around the building for any period.

10.2.2 Sample Collection

Meliponine specimens, honey and propolis samples were collected during the acquisition of colonies in the source area of origin and later after adapting to the PPBF. The collected samples were brought to the laboratory on the same day for morphological identification and further processing. Meliponine samples were stored in 95% ethanol and kept at 4 ° C for processing.

10.2.3 Species Identification of Meliponines and Specimen Preparation

Meliponine species were collected in separate vials. Three samples per specimen were captured to prepare these specimen pins. One container filled with isopropanol[66] and the meliponine is placed in the container to paralyze the meliponine. Pinned specimens were dried at 50C for 16 hours and then labelled. Following the method in Borror, D. J. and R. E. White 1970. A Field Guide to the Insects. The readied sample is stored in a clean container and placed in a silica gel to prevent moulds' growth.

10.2.4 Extraction and Linkage of Meliponines DNA

Meliponine dissection was performed under a stereomicroscope. The head and abdomen of the meliponine are pulled out and stored at -80 ° C. The left *thoracic* part of Meliponine extracted genomic DNA using the DNA easy Tissue extraction kit (QIAGEN Inc., USA). The DNA samples were then used as templates in the follow-up PCR using the primary point in Schedule 1. The PCR program ran following the sequence; initial denaturation for 2-5 minutes at 94-95 ° C followed by 34 cycles covering 1-minute denaturation at 94- 95 ° C; 30-60 when gilding at a temperature of 55-60 ° C and lengthening for 1 minute at a temperature of 72 ° C.

Schedule 1 shows the primary list used for species determination using the DNA barcode technique. PCR products were attenuated using the QIAquick Spin Kit (QIAGEN Inc., USA) and referred to using ABI3730xl.

Schedule 1. Primary list for determination of Meliponine species using DNA barcodes

---

[66] Editor's note: This is unfortunate. Good specimens for visual identification require collection without the addition of liquids, which often distort body parts and gum up hairs.

Barcode	Sequence	Reference
mitochondrial cytochrome C-oxidase subunit I (CO1)	SBEE-CO1-F: 5'-ATAATTTTTTTTATAGTTATAC-3' SBEE-CO1-R: 5'-GGAAAAAAAGTTATATTAACWC-3'	Ramirez et al. 2010
mitochondrial 16S rRNA (16S)	SBEE-16S-F: 5'-TTATTCACCTGTTTATCAAAACAT-3' SBEE-16S-R: 5'-TATAGATAGAAACCAATCT-3'	Cameron & Williams 2003
Arginine kinase (ArgK)	SBEE-ArgK-F: 5'- GACAGCAARTCTCTGCTGAAGAA-3' SBEE-ArgK-R: 5'- GGTYTTGGCATCGTTGTGGTAGATAC-3'	Kawakita et al. 2003
Polymerase II (PolII)	SBEE-PolII-F: 5'-AAYAARCCVGTYATGGGTATTGTRCA-3' SBEE-PolII-R: 5'-AGRTANGARTTCTCRACGAATCCTCT-3.'	Ramirez et al. 2010

10.2.5 DNA Reference Data Processing and Meliponine Phylogeny Analysis

The barcode reference data for each juxtaposed sample filtered for the quality of the using phred software (Ewing et al. 1998, Ewing & Green 1998), Then compiled with phrap software (http://www.phrap.org/phredphrapconsed.html) to become a complete bar code point and relate respectively. Completed reference barcodes for each sample collected following a collection of barcodes (COI, 16S, ArgK, PolII) and phylogenetic items were analysed using MEGA X filling (Kumar et al., 2018).10.2.6 Determination of Honey Metabolite Profile

10.2.5.1 Profile of Meliponine Honey Metabolite using 1D Nuclear Magnetic Resonance

The adjustment was carried out to the original method by Ohmenhauser et al. (2013) to minimize samples' use. In summary, a total of 250 mg of honey sample was dissolved in 1 mL of sterile ultrapure water and homogenized at 600 rpm, 20°C for 1 hour on a shaker. The insoluble material was removed from the homogenized sample utilizing a thumb blow at 10,000 rpm, 4°C, for 10 minutes. Four parts of the example mixed with 1 part of the pH 6.0 NMR buffer (1 M KH2PO4, 0.1% TMSP, 2mM NaN3 in D2O) and subjected to NMR spectroscopy under specific conditions - 1H NMR spectrum recorded at 300 K on the Bruker Ascend 700 MHz NMR Spectrometer (Bruker Biospin, Germany) using a 5mm three-resonance TCI Cryoprobe (frequency 700.4 MHz 1H).

10.2.6.2 Data Acquisition and Pre-processing

32 induction and four dummy effects of 65 k points were obtained with a spectrum width of 15,939 ppm, a receiver of 8 and a gain of 1.46 s. All NMR spectra were manually generated to correct baseline and calibrated by TSP signal at 0.0 ppm using Topspin 3.10.2.

10.2.6.3 Spectral Analysis

The NMR pre-processed data was transferred to Chenomx NMR Suite 8.1, which compares the detection of a known reference signal (TSP) with the signal derived from the associated reference containing the chemical shift and peak density. The change in each coat was identified based on the Chenomx database.

10.3 Results and Discussion

10.3.1 PPBF Meliponine Repository Development

A total of 6 Meliponines species have been transported from lowland areas to adapt to the environment in the PPBF. The ambient temperature lower than the source location causes sprawling (storage pots array) problems in the honey box, which triggers toadstool (fungus fruiting body) attacks and interferes with colony adaptation processes. Some of the colonies had also become weakened and abandoned. To overcome low temperatures (19 to 27 ° C), the topping box was customized using thicker wood and was equipped with an insulation layer to contain the heat. The insulation layer was placed around the logs to reduce the impression of a temperature drop on colony independence. *T. apicalis* and *L. terminata* colonies were able to adapt well after adding an insulation layer. More pots and honey were produced compared to other species.

a) The image in Topping box b) Colony image with topping box that was insulated.

Figure 61 a) Image in Topping box

Figure 60 b) Colony image with topping box insulated.

10.3.2 Sampling of meliponines for barcode analysis

Meliponine samples from the area around the PPBF and the new logs were brought to the terminal for species identification purposes using barcode DNA technology. Three primary molecular markers were used to identify the species obtained, namely 16S, CO1, and ArgK. DNA referencing carried out on Meliponines specimens at PPBF has identified unique species rarely found in lowland areas. Still, there is a conflict in the introduction of species morphologically versus using the 16S barcode method. Schedule 1 shows the introduction of Meliponines species found in the PPBF environment using morphometric techniques and 16S DNA barcodes techniques, where there is ID debris on BF02, BF03, BF04, BF05, BF06, BF07, BF08, BF10, BF11 BF12 and BF17. The three questionable species pairs were *Tetragonula geissleri* with *T. sapiens, T. laeviceps* with *T. sirindhornae* and *Tetragonilla collina* with *Tetragonilla fuscibasis*.

Schedule 2. Introduction of Meliponines species found in the PPBF area using morphometric techniques and 16S DNA barcode techniques

ID Specimen	Morphology	16s Barcode	ArgK Barcode
BF01	***H. bakeri***	***H. bakeri***	*L. terminata*
BF02	*T. geissleri*	*T. sapiens (97.2%)* *T. geissleri (96.8%)*	*T. geissleri*
BF03	*T. geissleri*	*T. sapiens (97.2%)* *T. geissleri (96.8%)*	*T. geissleri*
BF04	*T. laeviceps*	*T. sirindhornae (96.9%)* *T. laeviceps (96.6%)*	*T. laeviceps*
BF05	*T. geissleri*	*T. sapiens (97.2%)* *T. geissleri (96.8%)*	*T. geissleri*
BF06	*T. geissleri*	*T. sapiens (97.2%)* *T. geissleri (96.8%)*	-
BF07	*T. laeviceps*	*T. sirindhornae (94.9%)* *T. laeviceps (94.7%)*	*Tetragonula cf. pagdeni (98.2%)* *T. laeviceps (96.6%)*
BF08	*T. laeviceps*	*T. sirindhornae (94.9%)* *T. laeviceps (94.7%)*	*Tetragonula cf. pagdeni (98.2%)* *T. laeviceps (96.4%)*
BF09	***H. bakeri***	***H. bakeri***	***H. bakeri***
BF10	*T. geissleri*	*T. sapiens (97.2%)* *T. geissleri (96.8%)*	*T. geissleri*
BF11	*T. geissleri*	*T. sapiens (97.2%)* *T. geissleri (96.8%)*	*T. geissleri*
BF12	*T. geissleri*	*T. sapiens (97.2%)* *T. geissleri (96.8%)*	*T. geissleri*
BF13	***T. binghami***	***T. binghami***	***T. binghami***
BF14	***T. fuscibasis***	***T. fuscibasis***	-
BF15	***T. fuscibasis***	***T. fuscibasis***	*T. atripes (98%)* *T. fuscibasis (97.1%)*

BF16	***T. geissleri***	***T. geissleri***	***T. geissleri***
BF17	*T. collina*	*T. fuscibasis (99.2%)* *T. collina (98.1%)*	*T. collina*
BF18	***T. binghami***	***T. binghami***	***T. binghami***

Although high-quality reference data were obtained, some data were not comparable with data from the National Centre for Biotechnology Information (NCBI) database. The reference data for most meliponine species in Southeast Asia is not yet recorded in the NCBI public database. Reference and species identification data obtained from this project will be deposited into the NCBI database and Barcode of Life Data System (BOLD).

10.3.3 Principal phylogenetic analysis

The phylogenetic points of Maximum Likelihood for COI, 16S, ArgK and PolII barcodes are as follows:

a) COI

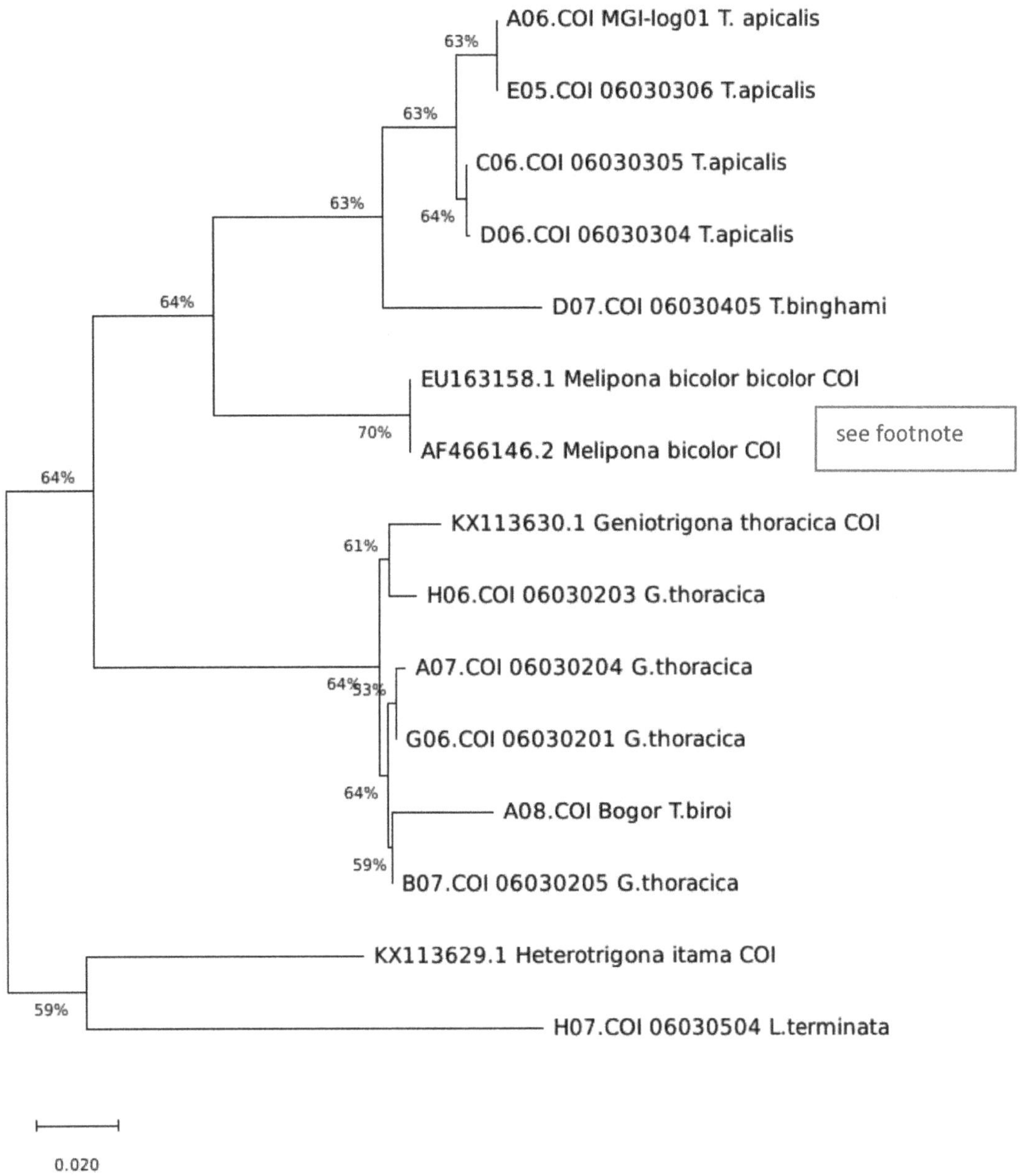

***Melipona bicolor*[67]

[67] It implies they evolved nested within two SE Asian bee clades, which cannot be true.

b) 16S

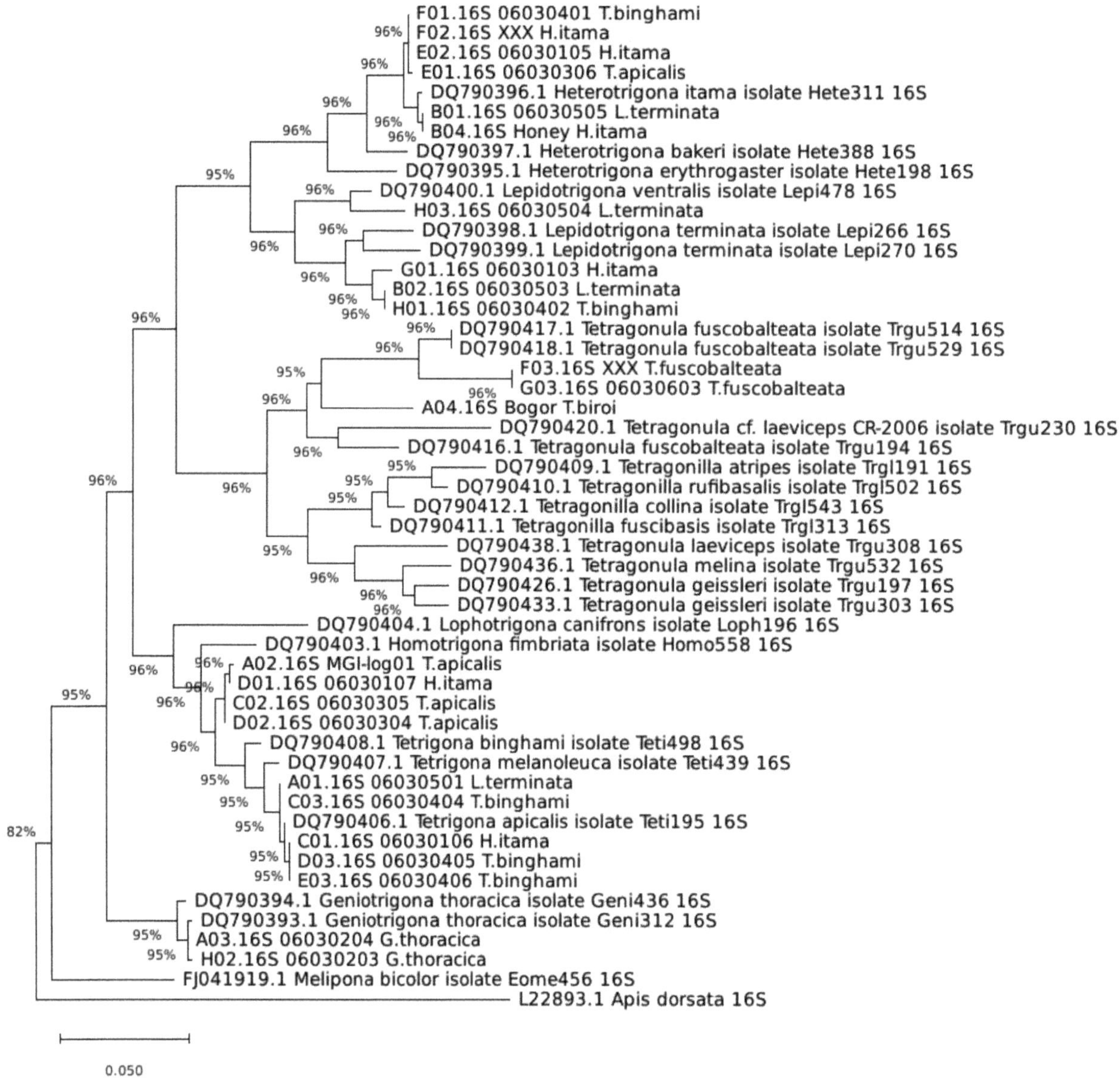
F01.16S 06030401 T.binghami
F02.16S XXX H.itama
E02.16S 06030105 H.itama
E01.16S 06030306 T.apicalis
DQ790396.1 Heterotrigona itama isolate Hete311 16S
B01.16S 06030505 L.terminata
B04.16S Honey H.itama
DQ790397.1 Heterotrigona bakeri isolate Hete388 16S
DQ790395.1 Heterotrigona erythrogaster isolate Hete198 16S
DQ790400.1 Lepidotrigona ventralis isolate Lepi478 16S
H03.16S 06030504 L.terminata
DQ790398.1 Lepidotrigona terminata isolate Lepi266 16S
DQ790399.1 Lepidotrigona terminata isolate Lepi270 16S
G01.16S 06030103 H.itama
B02.16S 06030503 L.terminata
H01.16S 06030402 T.binghami
DQ790417.1 Tetragonula fuscobalteata isolate Trgu514 16S
DQ790418.1 Tetragonula fuscobalteata isolate Trgu529 16S
F03.16S XXX T.fuscobalteata
G03.16S 06030603 T.fuscobalteata
A04.16S Bogor T.biroi
DQ790420.1 Tetragonula cf. laeviceps CR-2006 isolate Trgu230 16S
DQ790416.1 Tetragonula fuscobalteata isolate Trgu194 16S
DQ790409.1 Tetragonilla atripes isolate Trgl191 16S
DQ790410.1 Tetragonilla rufibasalis isolate Trgl502 16S
DQ790412.1 Tetragonilla collina isolate Trgl543 16S
DQ790411.1 Tetragonilla fuscibasis isolate Trgl313 16S
DQ790438.1 Tetragonula laeviceps isolate Trgu308 16S
DQ790436.1 Tetragonula melina isolate Trgu532 16S
DQ790426.1 Tetragonula geissleri isolate Trgu197 16S
DQ790433.1 Tetragonula geissleri isolate Trgu303 16S
DQ790404.1 Lophotrigona canifrons isolate Loph196 16S
DQ790403.1 Homotrigona fimbriata isolate Homo558 16S
A02.16S MGI-log01 T.apicalis
D01.16S 06030107 H.itama
C02.16S 06030305 T.apicalis
D02.16S 06030304 T.apicalis
DQ790408.1 Tetrigona binghami isolate Teti498 16S
DQ790407.1 Tetrigona melanoleuca isolate Teti439 16S
A01.16S 06030501 L.terminata
C03.16S 06030404 T.binghami
DQ790406.1 Tetrigona apicalis isolate Teti195 16S
C01.16S 06030106 H.itama
D03.16S 06030405 T.binghami
E03.16S 06030406 T.binghami
DQ790394.1 Geniotrigona thoracica isolate Geni436 16S
DQ790393.1 Geniotrigona thoracica isolate Geni312 16S
A03.16S 06030204 G.thoracica
H02.16S 06030203 G.thoracica
FJ041919.1 Melipona bicolor isolate Eome456 16S
L22893.1 Apis dorsata 16S
0.050

c) ArgK

G11.ArgK BF12
H10.ArgK BF05
F11.ArgK BF11
F10.ArgK BF03
E11.ArgK BF10
E10.ArgK BF02
A11.ArgK
DQ813097.1 Tetragonula melina isolate Trgu532 ArgK
DQ813079.1 Tetragonula geissleri isolate Trgu197 ArgK
DQ813091.1 Tetragonula geissleri isolate Trgu303 ArgK
C12.ArgK BF16
H12.ArgK MGI21
DQ813085.1 Tetragonula cf. laeviceps CR-2006 isolate Trgu230 ArgK
A10.ArgK Bogor T.biroi
DQ813078.1 Tetragonula fuscobalteata isolate Trgu194 ArgK
DQ813095.1 Tetragonula fuscobalteata isolate Trgu514 ArgK
DQ813096.1 Tetragonula fuscobalteata isolate Trgu529 ArgK
G09.ArgK 06030603 T.fuscobalteata
DQ813092.1 Tetragonula laeviceps isolate Trgu308 ArgK
G10.ArgK BF04
B11.ArgK BF07
C11.ArgK BF08
DQ813075.1 Tetragonilla collina isolate Trgl543 ArgK
DQ813073.1 Tetragonilla fuscibasis isolate Trgl313 ArgK
D12.ArgK BF17
DQ813072.1 Tetragonilla atripes isolate Trgl191 ArgK
DQ813074.1 Tetragonilla rufibasalis isolate Trgl502 ArgK
B12.ArgK BF15
G12.ArgK MGI20
DQ813050.1 Lophotrigona canifrons isolate Loph196 ArgK
E12.ArgK BF18 T.binghami
H11.ArgK BF13
DQ813065.1 Tetrigona apicalis isolate Teti195 ArgK
DQ813067.1 Tetrigona binghami isolate Teti498 ArgK
DQ813066.1 Tetrigona melanoleuca isolate Teti439 ArgK
C08.ArgK 06030305 T.apicalis
C09.ArgK 06030404 T.binghami
D09.ArgK 06030405 T.binghami
E09.ArgK 06030406 T.binghami
E07.ArgK 06030306 T.apicalis
F12.ArgK MGI19 T.binghami
DQ813045.1 Lepidotrigona terminata isolate Lepi266 ArgK
DQ813046.1 Lepidotrigona terminata isolate Lepi270 ArgK
B07.ArgK 06030505 L.terminata
A07.ArgK 06030501 L.terminata
DQ813047.1 Lepidotrigona ventralis isolate Lepi478 ArgK
H09.ArgK 06030504 L.terminata
B09.ArgK 06030205 G.thoracica
G08.ArgK 06030201 G.thoracica
DQ813038.1 Geniotrigona thoracica isolate Geni436 ArgK
DQ813037.1 Geniotrigona thoracica isolate Geni312 ArgK
A09.ArgK 06030204 G.thoracica
DQ813040.1 Heterotrigona erythrogaster isolate Hete198 ArgK
D10.ArgK BF01
D11.ArgK BF09
DQ813042.1 Heterotrigona bakeri isolate Hete388 ArgK
D07.ArgK 06030107 H.itama
DQ813041.1 Heterotrigona itama isolate Hete311 ArgK
C07.ArgK 06030106 H.itama
E08.ArgK 06030105 H.itama
F08.ArgK
G07.ArgK 06030103 H.itama
FJ042195.1 Melipona bicolor isolate Eome456 ArgK
AY267178.1 Apis dorsata ArgK
C10.ArgK Honey H.itama 2
0.10

d) PoIII

F05.PolII BF11
G05.PolII BF12
F04.PolII BF03
E05.PolII BF10
E04.PolII BF02
A05.PolII BF06
H04.PolII BF05
G04.PolII BF04
B05.PolII BF07
C05.PolII BF08
C06.PolII BF16
GU245411.1 Tetragonula carbonaria PolII
EU980086.1 Tetragonula fuscobalteata PolII
F03.PolII
A04.PolII Bogor T.biroi
A06.PolII BF14
D06.PolII BF17
B06.PolII BF15
G06.PolII MGI20
B03.PolII 06030205 G.thoracica
H02.PolII 06030203 G.thoracica
G02.PolII 06030201 G.thoracica
A03.PolII 06030204 G.thoracica
EU980077.1 Lepidotrigona ventralis PolII
H03.PolII 06030504 L.terminata
A01.PolII 06030501 L.terminata
B01.PolII 06030505 L.terminata
C01.PolII 06030106 H.itama
E02.PolII 06030105 H.itama
D01.PolII 06030107 H.itama
F02.PolII
G01.PolII 06030103 H.itama
D04.PolII BF01
D05.PolII BF09
EU162913.1 Melipona bicolor bicolor PolII
EU980083.1 Tetrigona binghami PolII
D03.PolII 06030405 T.binghami
F06.PolII MGI19 T.binghami
F01.PolII 06030401 T.binghami
E06.PolII BF18 T.binghami
E03.PolII 06030406 T.binghami
H05.PolII BF13
C03.PolII 06030404 T.binghami
E01.PolII 06030306 T.apicalis
D02.PolII 06030304 T.apicalis
A02.PolII MGI log01 T.apicalis
C02.PolII 06030305 T.apicalis
H01.PolII 06030402 T.binghami
0.050

10.3.4 Comparison of Meliponine Honey Metabolite Profiles before and after Adaptation to PPBF

The study of the metabolite profile of meliponine honey was carried out using 1H-Nuclear Magnetic Resonance (NMR) technology. The NMR platform application allows metabolite fingerprinting to be performed on different meliponine honey samples to identify unique and specific characteristics through the comparative spectrum and relative metabolite characterization using the Chenomx NMR Suite 8.1 database filling. The metabolite profile of meliponines honey obtained shows:

- the main sugar content of glucose, fructose, sucrose and maltose,
- amino acids such as phenylalanine, alanine, valine, isoleucine and tyrosine,
- organic acids such as acetic acid, lactic acid and formic acid, and
- ethanol at different relative concentrations.

The results of the previous analysis showed that there were differences in the content of lactic acid and formic acid in the honey of *T. apicalis* (log# 06030304), *T. binghami* (log# 06030404 and log# 06030406) and *L. terminata* (log# 06030501), where the formic acid content was higher. after adaptation versus before adaptation. It differs from lactic acid, where the content of these metabolites is lower after transformation than before transformation. 1H samples of *T. apicalis* spectra before and after adaptation - A noticeable difference can be seen in the circled area, representing the concentrations of formic acid (a), lactic acid (b) and ethanol (c).

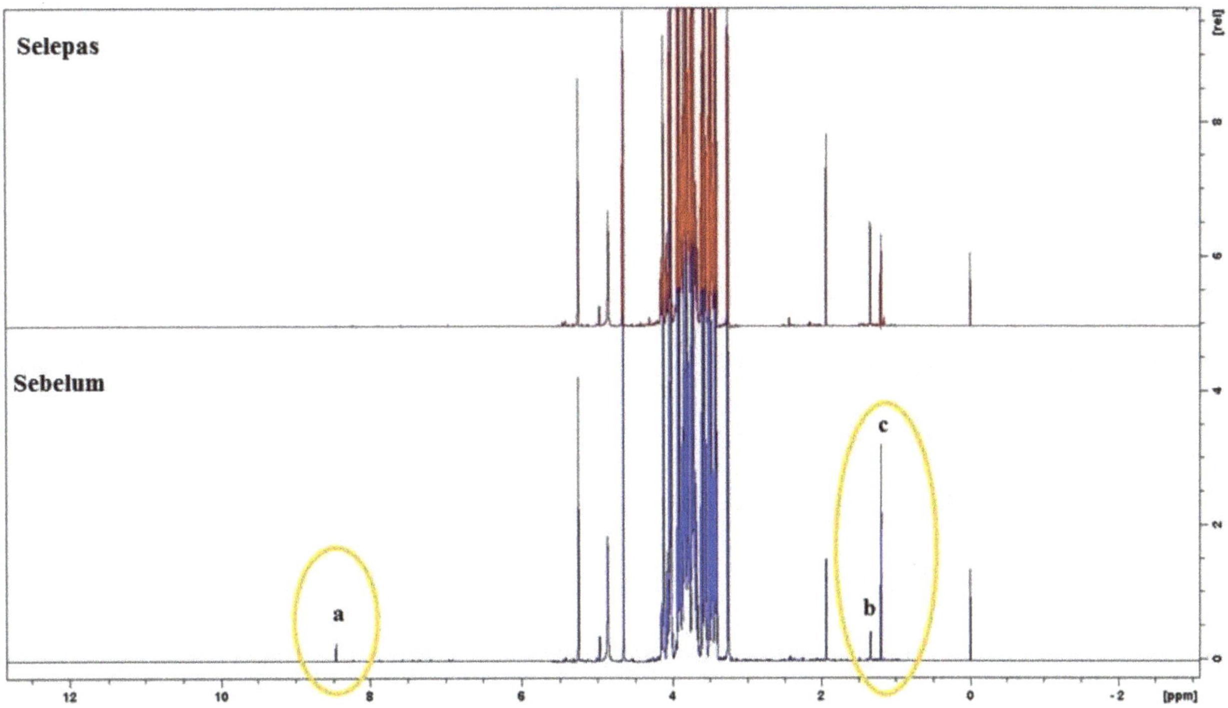

Figure 62 1H samples of T. apicalis spectra before and after adaptation

Changes in the two organic acids' content levels may be due to microbial flora's action to pressure ambient temperature changes. Formic acid plays a role in the biosynthesis of various compounds for energy metabolism and pathways related to the stress response. The process of reaching sugar also produces formic acids, which are involved in the phosphate oxidation process. Lactic acid is also a product of the activity of glycolysis metabolism. The increase in the concentration of lactic acid in honey may stem from the skin suppression system due to the metabolic process of acidosis of muscle tissues used at high frequencies for wing movement—excessive lactic acid associated with stress (Strachecka et al. 2019).

The resulting metabolite products are believed to be related to the physical activity of kelp during the adaptation period, in which transits from low soils with light hive boxes and hot temperatures, food sources are limited, and long distances of movement are applicable. Meliponine colonies' adaptation to elevated land at 1,200 meters above sea level is facilitated by forming insulated hive boxes and providing an environment with a good food source. The insulated hive box allows the gap between day and night temperature differences to be reduced and maintained at the optimum temperature level (25-28 ° C) for meliponines. Sources of resin, nectar, and debris found in abundance in the Bukit Fraser Investigation Centre's vicinity. However, analysis of the metabolite profile of meliponine honey after adaptation does not contain formic acid. Further studies were carried out to see the relationship between pressure and environmental factors on honey's metabolite profile.

10.4 Conclusion

Through the decision, the three main objectives were successfully achieved.

10.5 References

Momose, K., Yumoto, T., Natamitsu, T., Kato, M., Nagamasu, H., Sakai, S., Harrison, R.D., Itioka, T., Hamid, A. A. & Inoue, T. 1998. Pollination biology in a lowland Dipterocarp forest in Sarawak, Malaysia. I. Characteristics of the plant-pollinator community in a lowland dipterocarp forest. American Journal of Botany, 85(10): 1477-1501.

Nagamitsu, T., Momose, K., Inoue, T. & Roubik, D.W., 1999. Preference in flower visits and partitioning in pollen diets of stingless bees in an Asian tropical rain forest. Research on Population Ecology, 41: 195-202.

Ramalho, M., Giannini, T. C., Malagodibraga, K. S., & Imperatriz-Fonseca, V.L. 1994. Pollen harvest by stingless bee foragers (Hymenoptera, Apidae, Meliponinae). Grana, 33:239–44

Heard, T. A., 1999. The role of stingless bees in crop pollination. Annual Review of Entomology, 44:183–206

Salim, H. M. W., Dzulkiply, A. D., Harrison, R.D., Fletcher, C., Kassim, AR, & Potts, M.D., 2012. Stingless bee (Hymenoptera: Apidae: Meliponini) diversity in dipterocarp forest reserves in Peninsular Malaysia. The Raffles Bulletin of Zoology, 60(1): 213-219.

Cameron, S.A. & Williams, P. H., 2003. Phylogeny of bumblebees in the New World subgenus Fervidobombus (Hymenoptera: Apidae): congruence of molecular and morphological data. Molecular Phylogenetics and Evolution 28:552-563

Kawakita, A., Sota, T., Ascher, J. S., Ito, M., Tanak, H., & Kato, M. 2003. Evolution and phylogenetic utility of alignment gaps within intron sequences of three nuclear genes in bumblebees (Bombus). Molecular Biology & Evolution 20: 87-92

Sulaiman, N. 2018. Pengesahan spesies Tetragonula geissleri melalui kajian morfometrik dan kajian biomolekul di Kawasan tanah tinggi dan tanah rendah. Tesis Sarjanamuda Sains, FST, UKM.

Kumar S., Stecher G., Li M., Knyaz C., and Tamura K. (2018). MEGA X: Molecular Evolutionary Genetics Analysis across computing platforms. Molecular Biology and Evolution 35:1547-1549.

Ewing B, Green P. (1998). Base-calling of automated sequencer traces using phred. II. Error probabilities. Genome Research 8:186-194

Ewing B, Hillier L, Wendl M, Green P. (1998). Base-calling of automated sequencer traces using phred. I. Accuracy assessment. Genome Research 8:175-185.

Ramírez SR, C Nieh, TB Quental, DW Roubik, VL Imperatriz-Fonseca, NE Pierce (2010) A molecular phylogeny of the stingless bee genus Melipona (Hymenoptera: Meliponini). Molecular Phylogenetics and Evolution 56(2): 519-525

Chapter 11

11.0 Meliponine Repository at Malaysia Genome Institute (MGI)

By Mohd. Razif Mamat

Figure 60 Meliponine Repository at the Malaysia Genome Institute (MGI)

A premier network-based institute in genome research for knowledge generation, innovation and technology transfer for economic development

11.0.1 Introduction

In 2016, with the establishment of Akademi Kelulut Malaysia, we found that many affiliates and AKM participants had in possession varied meliponine species with questionable foraging sources. Many of them had unidentified species with little know foraging habits. Many meliponiculture efforts were failing due to that lack of knowledge. Simultaneously, AKM collaborated with MGI to ascertain these answers in species ID and foraging habits.

MGI had in its compound a wide variety of wild vegetation (reserved for future expansion) and decided to establish a repository of different species within the MGI compound to be able to scrutinize within reach and reasonable time.

What resulted was a confluence of donated hives from AKM members for a shared knowledge base. Many beekeepers contributed more hives who could not continue their meliponiculture of peculiar and rare species failing in their possession from that initial step. In order to save and learn about those problematic types, we moved them to MGI for ultimate conservation. Hence, evolved the MGI

Meliponine repository. Seeing the positivity in this effort, some learning institutions, with collaboration with research institutions and the agriculture department, followed suit and established similar warehouses in other locations.

For both MGI and AKM, the repository has provided easy and quick access to sampling and diverse research in Meliponine species ID and genomics, studies in foraging habits, meliponary management and predator peculiarities. Research in Meliponine honey by MGI has resulted in much knowledge now, especially in metabolite profiling.

11.1 MGI Core Facilities

- Set up in early 2016
- Collaboration MGI and Akademi Meliponines Malaysia
- To date, have 17 identified species and more than ten species not yet been identified

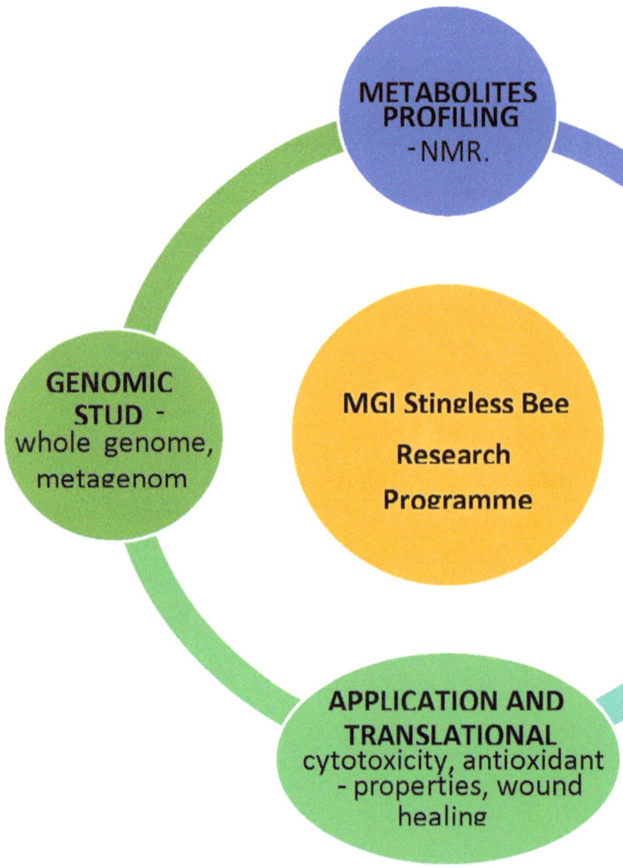

Figure 61 MGI Core Facilities

Sequencing

High throughput sequencing

Illumina HiSeq 2000, MiSeq
8 ABI Sequencers 3730XL

Genome Computing Centre

Genome and Structural Informatics
SGI Altix 4700 Super Cluster, SGI XE 1200 Cluster, SGI
Infinite Storage 350,
Software - CLC Genomics Workbench, Velvet, Discovery Studio, Staden & Phrap package

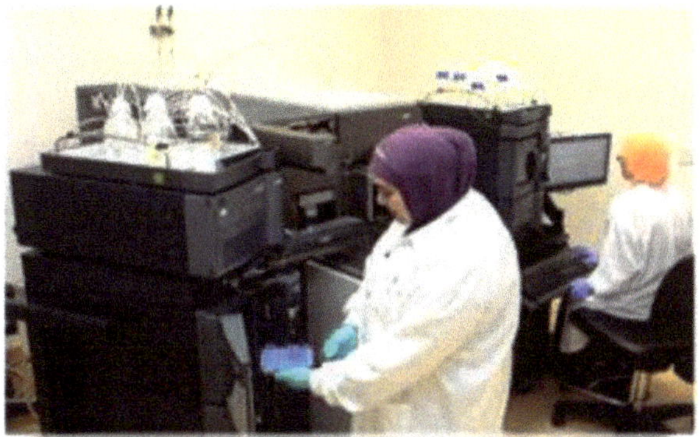

Proteomics

High throughput characterization and identification of large proteins and peptide mixtures
Waters 2D-LC-MS/MS,
Protein Expression System Informatics,
INSIGHT 2.0, MassLynx v 4.1

Protein Engineering & Bioprocess

Expression and purification of prokaryotic and eukaryotic proteins
INFORS Bioreactors, AKTA PRIME Purification Systems, Cross flow filtration Systems

Structural Biology & Biophysics

Wide information of structural, dynamics and interactions of biomolecules
Bruker Ascend 700 MHz NMR

Structural Biology & Biophysics

Measuring the position and intensity of the diffracted X-rays of protein crystal to generate data for 3D structure development
Single X-Ray Diffractor System

11.2 List of species at repository

1.Tetrigona binghami
2.Tetrigona apicalis
3.Tetrigona melanoleuca
4.Homotrigona fimbriata
5.Lophotrigona canifrons
6.Tetragonula laeviceps
7.Tetragonula fuscobalteata
8.Tetragonula atripes
9.Tetragonula cf. *geissleri*
10.Tetragonula cf. *minor*
11.Tetragonula testaceitarsis
12.Geniotrigona thoracica
13.Lepidotrigona doipaensis
14.Lepidotrigona terminate
15.Lepidotrigona cf. *ventralis*
16.Heterotrigona itama
17.Heterotrigona erythrogastra
More than ten species have not yet to identified

11.2.1 List of depositors

1.Mr. Badrul Mohd Shah
2.Mr. Wak Abdul Aziz Bakri
3.AHJ & Dr. Ibrahim Bin Shuib
4.Mr. Hendra Ja
5.Mr. Jeffry Iskandar Foo
6.Ms. Nur Raziana Ramli
7.Prof Nik Marzuki
8.Mr. El Ilmy
9.Mr. Hadi Mua'fi Taufik
10.Tengku Zulkifli
11.Haji Mohd Jolkapli
12.Ms. Ana dan Ms. Azy (UKM)
13.Ms. Halimah Alias
14.Mr. Zanari
15. Tn Aminaddin (Tati Agro Farm).
16. Mr. Mohamad Amiruddin Bin Othman
17.Haji Talib Md Seh

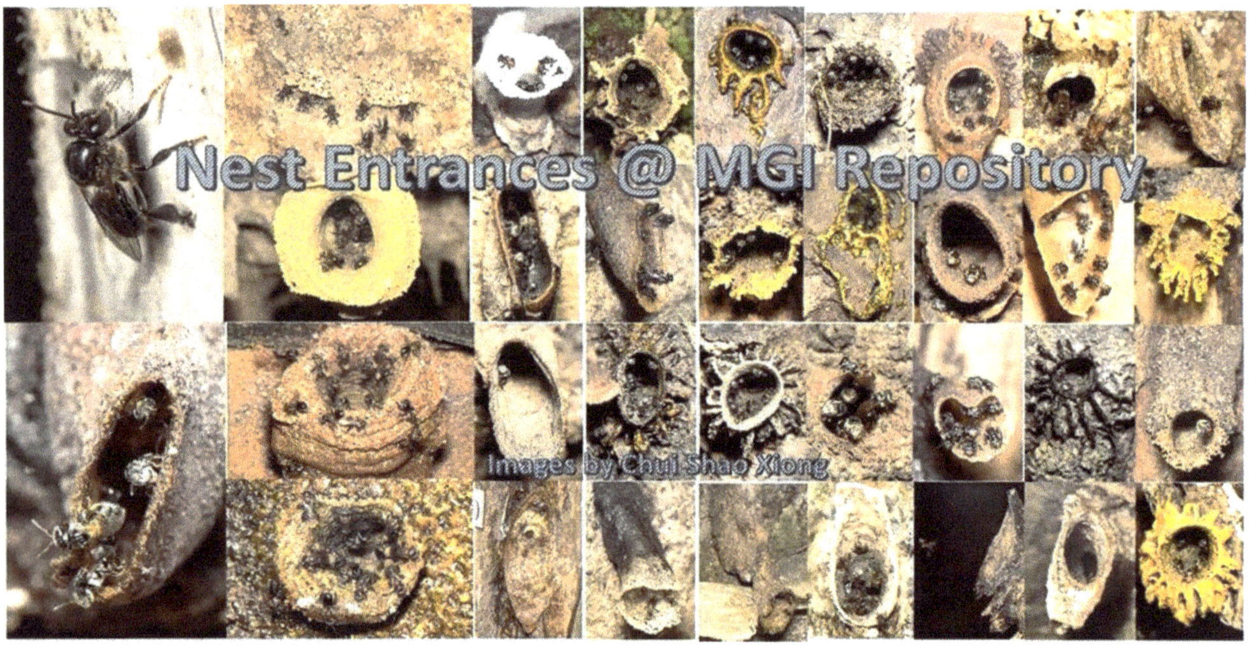

Figure 62 Best Entrances at MGI Repository Photo by Chui Xiao Siong

11.4 Importance of Meliponine Repository

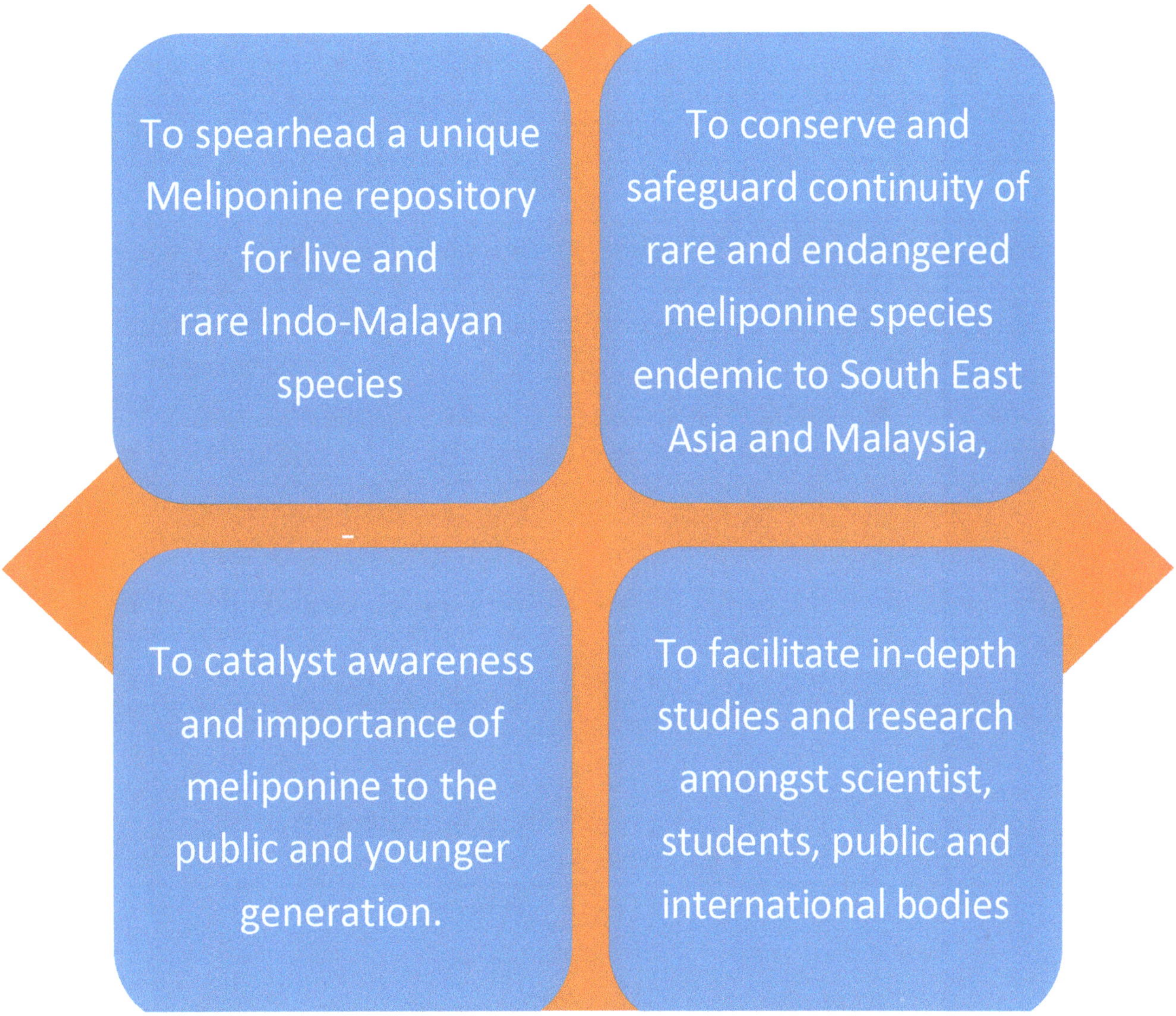

Figure 63 Importance of Meliponine Repository

Heterotrigona itama
Fructose ....(<50%) [40.6%]
Glucose ....(>50%)[(56.2%]
Sucrose ....(<1%)[0.7%]
Maltose ....(>1%)[2.5%]
Other metabolites:
Phenylalanine, Alanine, Valine, Tyrosine, Acetat
Lactate, **Trigonelline, Ethanol**

LOCATION	*Unknown*
Pasar Cameron Highland, Pahang (Madu Putih),	GFSM Mixture: Fructose ....<50% Glucose ....>50% Sucrose ....nil Maltose ....nil May contain trace of: Arabinose, Betaine, Cellobiose, Citrate, Ethylene glycol, Glycine, Glycolate, Guanidinoacetate, Ribose
Pasar Cameron Highland, Pahang (Madu Perang), unknown origin	GFSM Mixture: Fructose ....<50% Glucose ....>50% Sucrose ....<10% Maltose ....<50% May contain trace of: Ethanol, Glycolate, Guanidinoacetate, Xylitol
Kawalan Madu Tiruan Genom	GFSM Mixture: Fructose ....<50% Glucose ....<50% Sucrose ....>10% Maltose ....nil May contain trace of: Citrate

Figure 65 H. itama Metabolite Profile

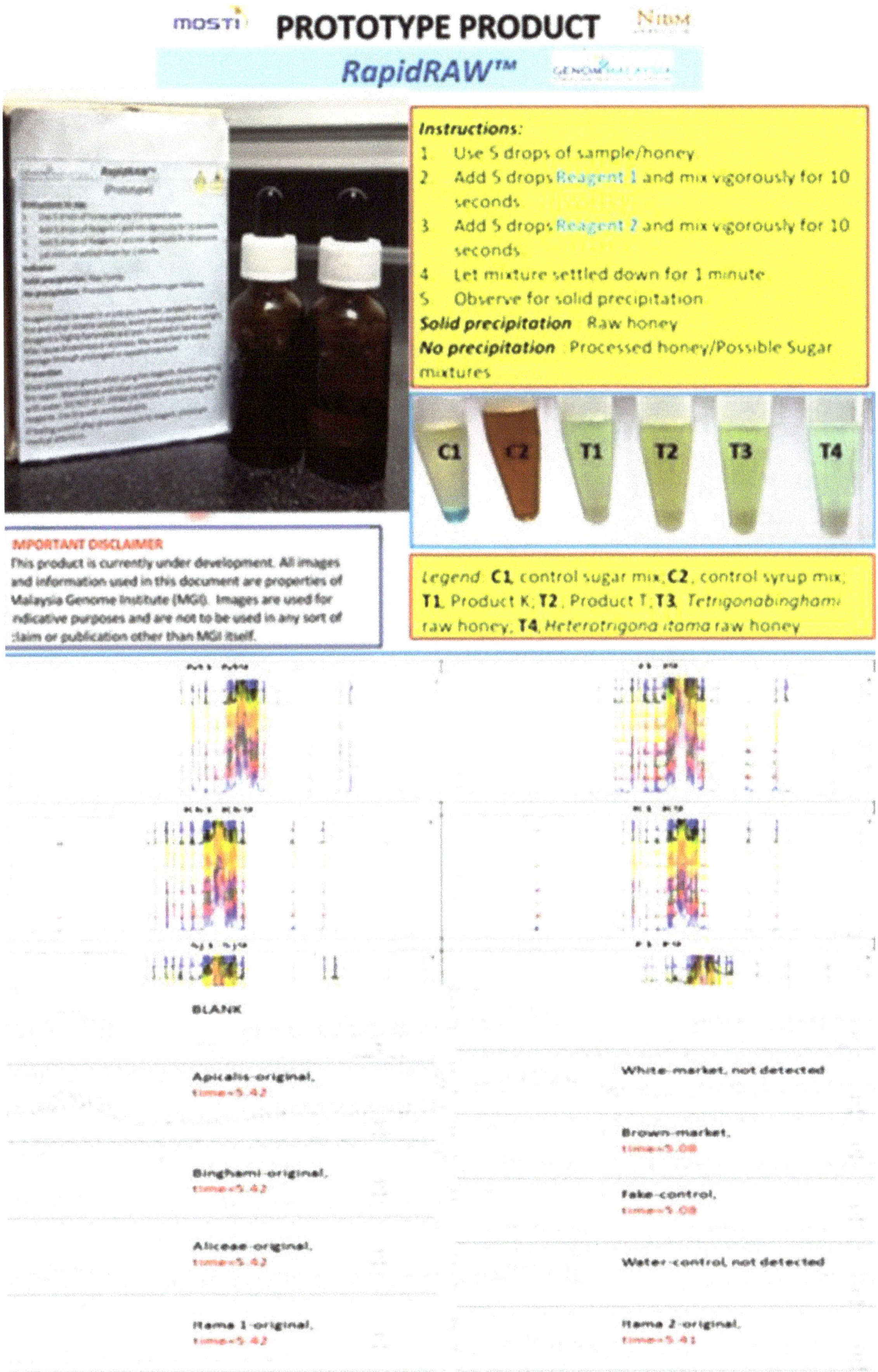

Figure 66 RapidRAW product Prototype poster

11.6 RapidRAW Reaction Spectrum

11.6.1 Previous packing

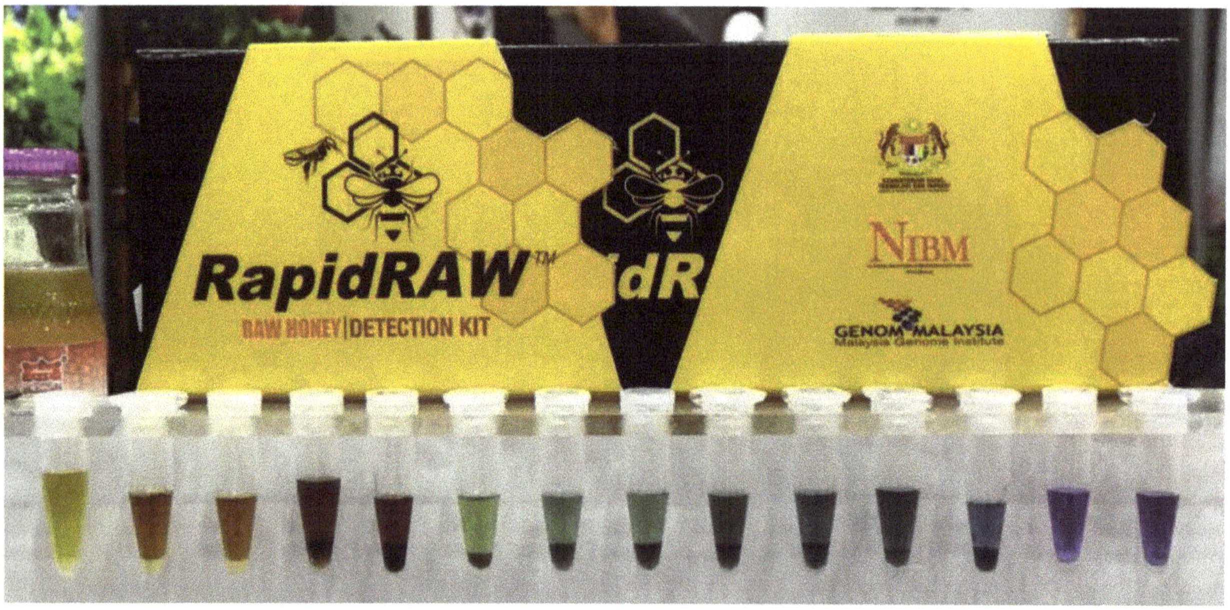

11.6.2 New packing

Figure 67 RapidRAW Reaction Spectrum - Screening Test Kit; Previous packing above and New packing below

11.7 SDS PAGE GEL

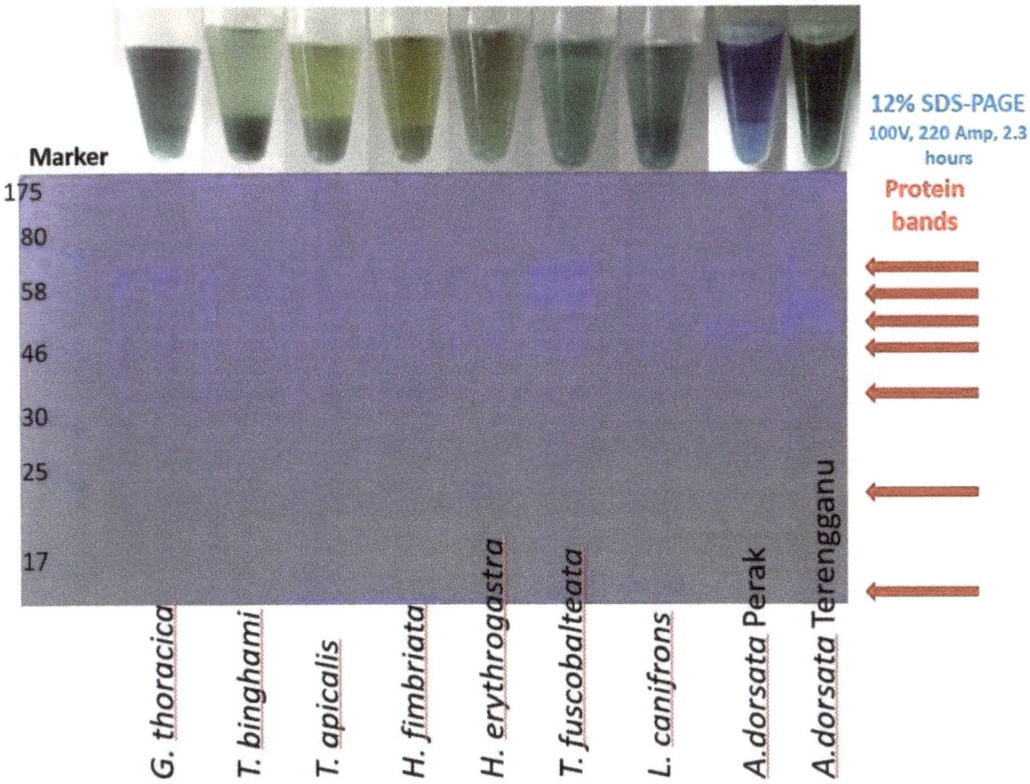

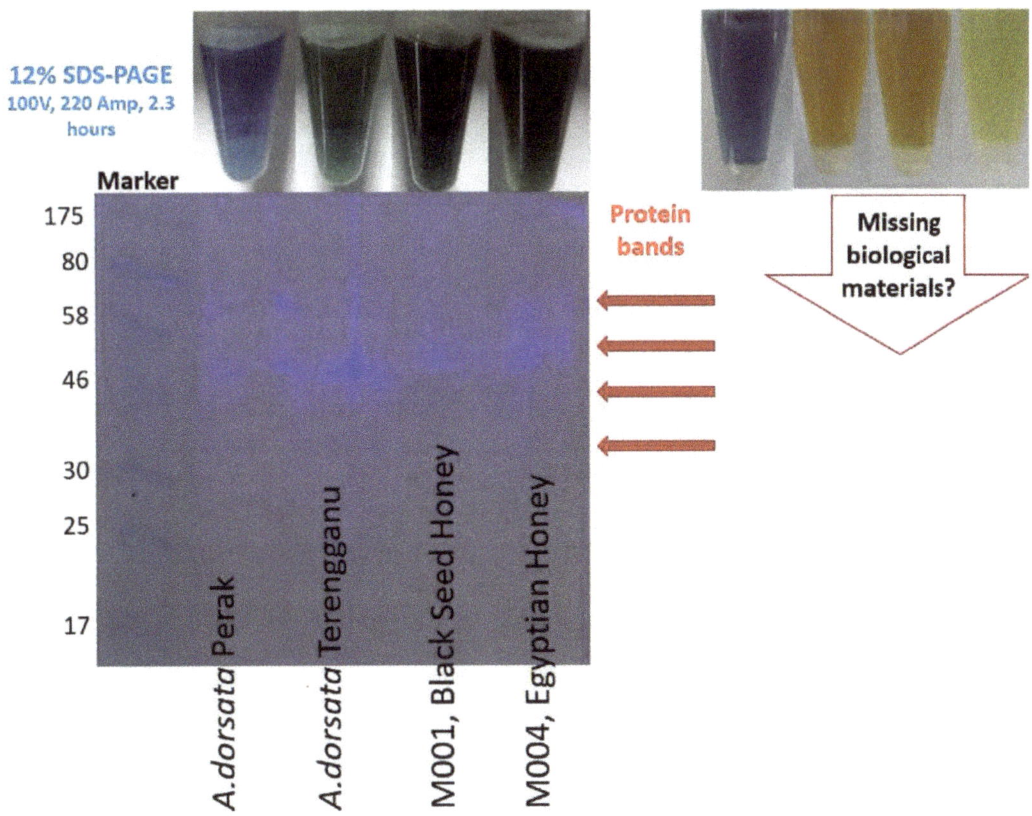

Figure 68 SDS PAGE GEL

11.8 Application of DNA Barcoding in Identifying Stingless Bee Colonies - MSI 16046

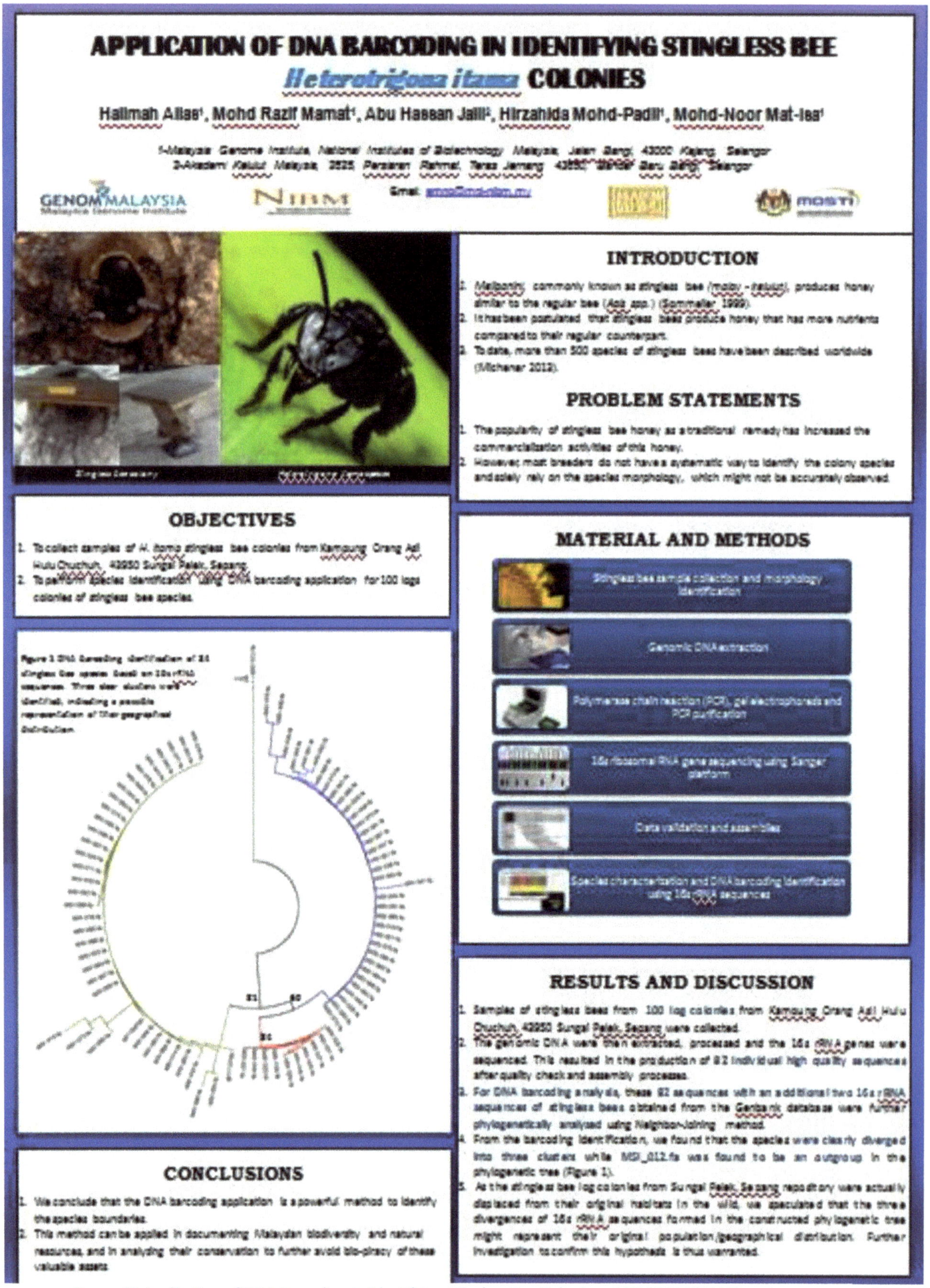

Figure 69 Application of DNA Barcoding in Identifying Stingless Bee Colonies - MSI 16046 (2016) poster (2016)

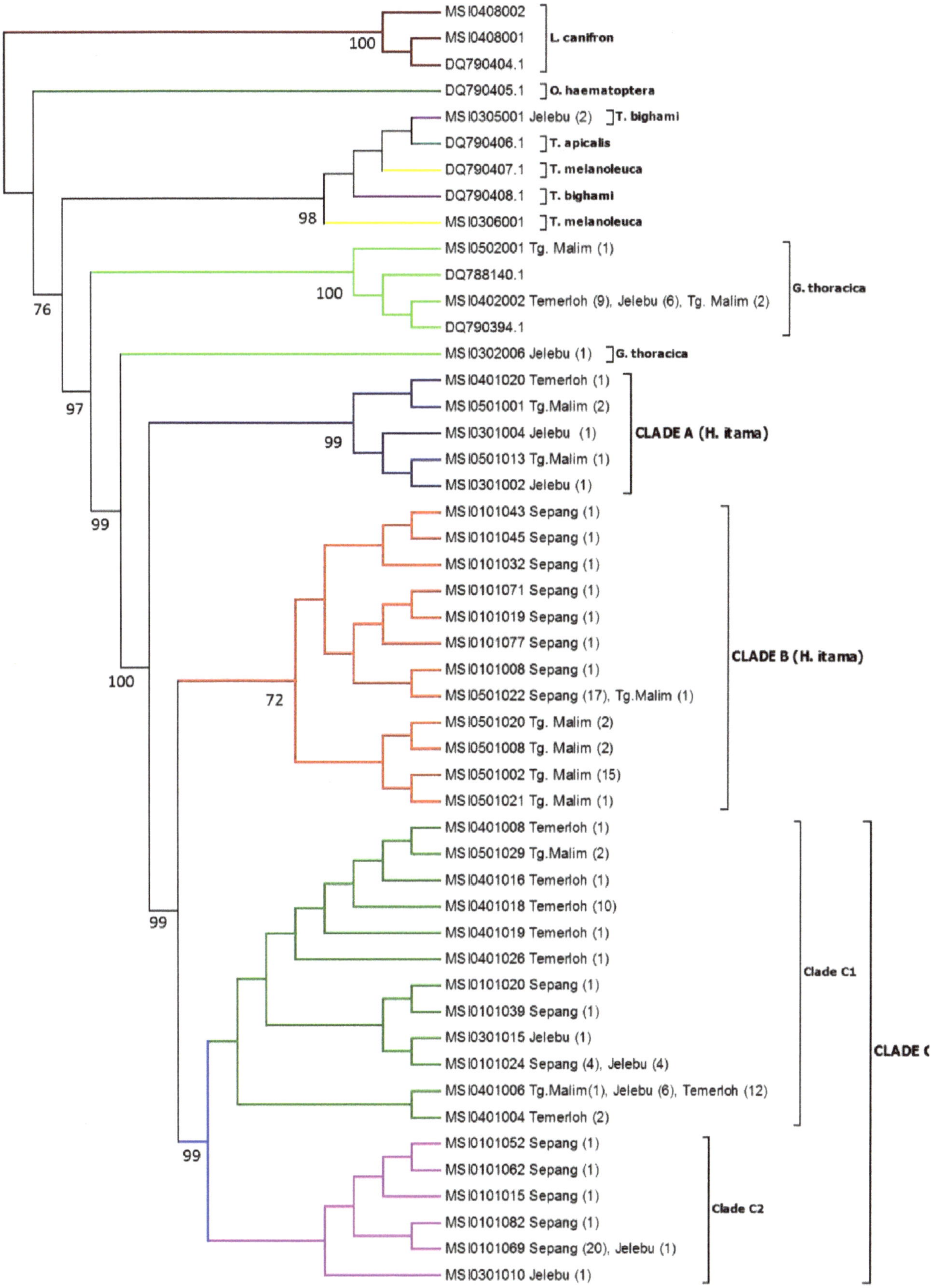

Figure 70 Phylogeny tree

11.9 DNA Barcoding

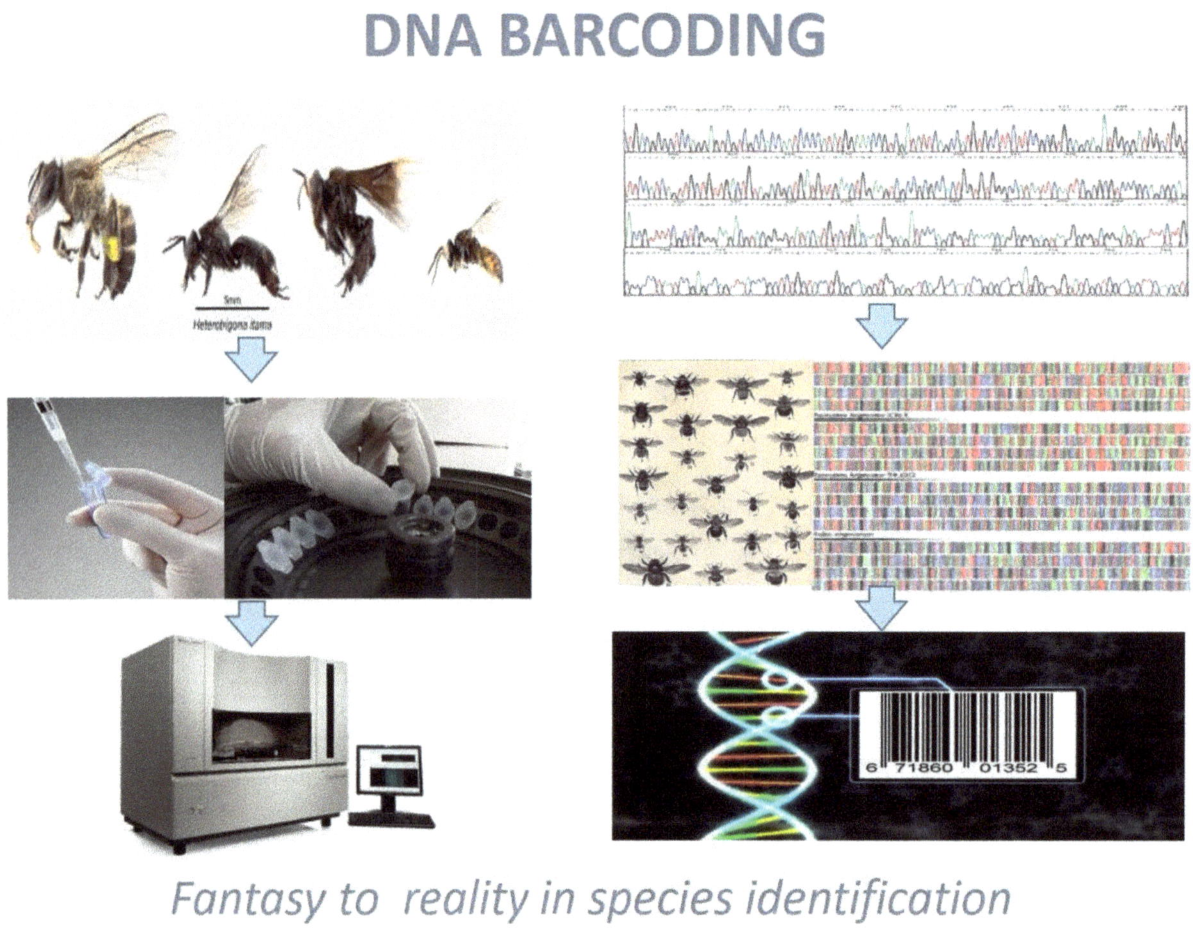

Figure 71 DNA Barcoding

GENOM MALAYSIA
Malaysia Genome Institute

National Institutes of Biotechnology Malaysia **Not-for-Profit CLG**

Chapter 12

12.0 Meliponiculture & Nature Therapy for Learning Disorders Kid@Ukm Fraser Hill Research Centre

Mohd Razif Mamat[1*], Azira Muhammad[1], Norela Sulaiman[2], Mohd. Afiq Aizat Jauhari[2], Halimah Alias[1], Nor Azfa Johari[1], Mohd Noor Mat Isa[1], Nazahiyah Ahmad Rodzli[1], Bibi Syahila Mohd Sohi[1], Wan Khairul Ikhsan Wan Seman[1], Mervynn Koon[3] & Jocye Tan[3],

[1]Malaysia Genome Institute, National Institutes of Biotechnology Malaysia, Ministry Science, Technology and Inovation (MOSTI), Jalan Bangi, 43000 Kajang, Selangor

[2]Department of Biological Sciences and Biotechnology; Fraser`s Hill Research Centre, Faculty Science and Technology, 43600 Universiti Kebangsaan Malaysia, Bangi, Selangor

[3]Andante Education Department, Andante International Foundation, 63 Jalan Ecohill 3/2C Setia Ecohill, 43500 Semenyih Selangor

Correspondence: razif@nibm.my; mrazifmamat@gmail.com

12.1 Abstract

Learning in a natural environment has proven time and again to boost the speed of knowledge transfer. In this study, we assessed nature therapy for children with learning disorders at UKM Fraser's Hill Research Centre. A team of six boys of the ages between 12 to 17 years old with a history of Autism, Dyslexia, Attention Deficit Hyperactive Disorder (ADHD), Oppositional Defiant Disorder (ODD), challenging behaviours, and non-conventional learners were involved in a 3-day program. Series of activities and workshops were organised such as Stingless Bee Presentation Session, Pitcher Plant Tour, Papermaking Session, Terrarium-making Session, Soap-making Session, Light Trap-making Session, Nature and Insect Observation and Jungle Trekking. Safety was highly prioritised, and the program was well adjusted to weather and time. The boys enjoyed the whole program and looked forward to another similar program. They were hungry for more knowledge and experience with the natural environment.

Keywords: Nature Therapy, Learning Disorder, Stingless Bee

12.2 Introduction

Autism Spectrum Disorder (ASD) is a complex developmental disorder associated with symptoms that include "persistent deficits in social communication and social interaction across multiple contexts" and "restricted, repetitive patterns of behaviour, interests, or activities. Oppositional defiant disorder (ODD) is a type of behaviour disorder. Children with ODD are uncooperative, defiant, and hostile toward peers, parents, teachers, and other authority figures. Attention-deficit/hyperactivity disorder (ADHD) is the most common neurobehavioral disorder of childhood. According to the Centres for Disease Control and Prevention: "if untreated, a person with ASD, ODD & ADHD will struggle with impairments in crucial areas of life, including relationships with peers and family members, and performance at school or work. Unfortunately, current ASD, ODD & ADHD treatments fall far short of ideal, offering only limited relief from symptoms and often involving serious side effects.

Learning in a natural environment has proven time and again to boost the speed of learning. Integration of beekeeping as a part of learning with nature should be a part of the curriculum for children with disabilities and all schools and learning centres. Nutritional properties of pollen and the regulating metabolic processes are used, among others, in children's lack of appetite, developmental delay and malnutrition of children and adults.

On the other hand, adaptogenic properties can also increase the resistance to harmful physical, chemical, and biological factors. Increasing physical fitness in excessive physical burden, affecting the central nervous system by improving brain functions, such as memory, learning, comprehending, thinking, and concentration ability, and increasing the immune system strengths against infection en route boosting the immunological system.

To implement meliponiculture as therapy, we can introduce a hive to homes, schools and public parks (provided all the precautions and safety factors are taken into consideration), education and public awareness. Working together with authorities and communities will make sure all the planning is successful implement.

This article reports the preliminary experiment to explore a possible treatment for ASD, ODD & ADHD through nature. Supposing controlled experiments and clinical trials bear out this potential. In that case, such natural treatments promise to supplement current approaches to managing ASD, ODD & ADHD, with the advantages of being widely accessible, inexpensive, not stigmatising, and free of side effects.

A team of six boys of the ages between 12 to 17 years old with a history of Autism, Dyslexia, Attention Deficit Hyperactive Disorder (ADHD), Oppositional Defiant Disorder (ODD), challenging behaviours, and non-conventional learners were involved in a 3-day program at UKM Fraser's Hill Research Centre. These teenagers were treated and closely monitored by Dr. Mervynn Koon. The programme was designed to give exposure to nature and teach the boys specific knowledge and skill. After the end of the session, we hope the children will learn something from nature and improve their lack of learning ability.

12.3 Materials and Methods

12.3.1 The Program Itinerary

DAY 1 (25, July 2019)	DAY 2 (26, July 2019)	DAY 3 (27, July 2019)
Arrival Tent Setup Teatime Pitcher Plant Tour Wash Up Prep for Night Presentation BBQ Dinner Presentation Session (Topic: Meliponary) Feedback Lights Out	Morning Call Breakfast Paper-making Session Jungle Walk (for the material hunt) Lunch Terrarium-making Session (with hunted materials) Soap-making Session Mandi Sungai at Air Terjun Dinner Light Trap Setup Jungle Trekking, Nature and Insect Observation Light Trap Review Lights Out	Morning Call Final Light Trap Review Breakfast Dismantling and packing up Closing Ceremony Group Photo Session Team Heads Back

12.3.2 Event highlights in the program

#	SESSION	FOCAL AREA
1	Pitcher Plant Tour	Nature, Plant
2	Meliponary Presentation Session	Entomology and Meliponary
3	Paper-making Session	Green Living
4	Terrarium-making Session	Plant, Nature, Green Living
5	Soap-making Session	Green Living
6	Light Trap-making Session and Review	Nature, Entomology, Science (observation skills)
7	Jungle Trekking	Nature, Plant, Entomology, Science (observation skills), Outdoor Skills
8	Nature and Insect Observation	Nature, Entomology, Science (observation skills)

12.3.3 Facilitators

#	Main Personnel	Information	Role In The Program
1	Mohd Razif Mamat	Scientist Translational Research & Prototype Development Centre	Overall Facilitator, Green Living Facilitator
2	Mohd. Afiq Aizat Juhari	Researcher Ketua Pusat Penyelidikan Bukit Fraser	Plant, nature, Entomology Facilitator
3	Mohd Noor Mat Isa	Scientist Genome Computing Centre	Program Facilitator
4	Wan Mohd Khairul Ikhsan Wan Seman	Scientist Protein Expression and Purification	Program Facilitator
5	Dr. Nazahiyah Ahmad Rodzli	Scientist Structural and Applied Genomics	Program Facilitator

6	Halimah Alias	Scientist Genome Sequencing	Program Facilitator
7	Bibi Syahila Md Sohi	Technologist Lab Services	Program Facilitator
8	David Sya	Outdoor Instructor Khootz Outdoor Ventures	Outdoor Facilitator
9	Joshua Chin	Plant Conservationist Preservation and Conservation Dept Andante Foundation	Outdoor Facilitator

12.3.4 Participants

A team of six boys of ages 12 to 17 from Khootz Outdoor Ventures was sent for this program. Among the many conditions once faced by the boys include; Autism, Dyslexia, Attention Deficit Hyperactive Disorder (ADHD), Oppositional Defiant Disorder (ODD), challenging behaviours and non-conventional learners. These boys have been learning and living in an outdoor natural environment for the past months, to which the degree of their conditions remarkably lessened. They are now in a stable phase where they can normally interact with their peers, adults around them and external people. This is their first time venturing to Bukit Fraser and utilising their basic knowledge and skills to live there.

12.4 Results and Discussion

12.4.1 Feedback Facilitators

The facilitators from Andante and Khootz saw a potential positive impact by scientists and teachers working together in such a venue (environment) for such micro workshops For the boys. There were astounding remarks of working with the MGI and UKM Facilitators as a team. Safety was well-considered, and with quick thinking, the program was well adjusted to weather and time. The program is not far off from what Khootz has been delivering. On the other hand, our facilitators believe there could be better event coordination from all sides for future sessions. They also noted that it would be good to have a brief introduction session amongst the facilitators (i.e. name, the designation, etc.). Other suggestions include a brief session on genetics and apiary, botany, ecology, water and soil. The goodies packs received were very useful. No major incidents took place, or major issues were highlighted.

12.4.2 Feedback Participants

The boys enjoyed the whole program and had much to look forward to another programme in the future. They were hungry for more knowledge and experience with the ecology found there. They loved the trekking and sessions. BBQ dinner was great. This was their first time running a presentation in front of a team of adults, all of who were new and unfamiliar to them. Upon reflection, this was their overcoming hurdle to public speaking and subsequently realising the need to learn more about the subject matter. The feedback given seemed to serve as a guiding point. They were excited over all the hands-on activities that were conducted on their second day. Participants were glad to each receive a goodie pack. They requested a better in-depth introduction to the personnel and sessions. When asked if they wished to return, the answer came back 'yes' with a big smile. No major incidents took place, or major issues were emphasized.

12.4.3 Questionnaire (Post Event) for Feedback and Evaluation

10 Questions Asked and Participant Answers

1. What did you remember most about the program?
    - Night jungle walk/hike, papermaking, bar soap-making, BBQ, making insect trap & observations
    - Spider (insect) observation
    - Walking nature trails
    - Honey harvesting (Stingless bee), jungle trekking and night walk
    - Waterfall
    - Soap-making, afternoon and evening hikes (nature and insect observations)

2. How much did you enjoy learning in the program? (on the scale of 1-10)

    Only two persons out of the whole group rated their enjoyment at 9/10. The others gave a perfect 10/10.

3. Describe your observations and feedback about the environment and facilities of the program.

Generally, everyone described their observations and appreciation of being surrounded by fresh air, cool air and a natural environment. They noted that the facilities are clean and tidy.

4. What did you learn most out of the program?
    - Soap-making (same answer from 3 persons)
    - Tarantula, insects and bugs
    - Stingless bee identification
    - Stingless bees and the ecosystem

5. If given another chance to attend the program, what would you like to have experience?
    - Night hike, swimming, searching for wildlife
    - More on tarantulas and insects
    - Birding, photography, jungle craft, wildlife and ecology, plants and seeds
    - Watch for rare animals and precious plants at Fraser's Hill, the study of animals, DIY activities
    - Learn more about animals and strawberry farms nearby
    - Bird watching, more trails and collect more specimens

6. Did you understand the content of the program? (e.g. the sessions, instructions, information, etc.)
    - Not all (especially Bahasa, and it was spoken too fast)
    - Yes, understanding the main points but couldn't pick up profound meanings
    - Yes, except those technical terms in Bahasa
    - The others replied, 'yes.'

7. How were the facilitators in the program?
    - Good, kind, polite and helpful
    - Kind and very humble
    - Friendly and helpful
    - Kind, thoughtful, helpful and patient in explaining words
    - Friendly and willing to teach
    - Kind and polite

8. How was the accommodation of the program?
    - Accommodation is portable (tent), neat, clean and tidy
    - Memorable

- Clean, tidy and comfortable
- Cooling, peaceful and quiet
- Clean, tidy, orderly, partly sheltered and comfortable

9. What are your thoughts about the length of the program?
    - Not enough, hope can stay longer
    - Length too short
    - Prefer months
    - To rush and activities done in a rush. Look forward to a week program
    - I think four days, three nights
    - Too short

10. Why would you encourage someone your age to attend the program?
    - Because it is far from the city area, the air is fresh and the right place to visit to learn more things there
    - Because of the potential transformation, the program can help to do for people my age and to encourage young people my age to devote their lives to nature
    - To promote a passion for nature, raise awareness of the environment and share the enjoyment with others
    - So that they can learn more about rare and precious plants and animals that we have in Malaysia
    - Because many people have not been given a chance to be in this kind of learning environment
    - Because many people my age do not have a chance to go outdoors to experience what I feel outdoors

12.5 Compilation of pictures

Figure 72 Fraser Hill Nature Camp Photo Gallery

Figure 73 Nature & Science at Fraser's Hill

12.6 Conclusion

Based on the participants' overall feedback, they acknowledged that they learned a lot from this programme. The programme made learning comfortable for them and learning no longer posed a challenge for those having conditions. Therefore, the programme had a therapeutic effect on them. On top of that, learning in a natural environment was key to improving their ability to learn. Real-life flora and fauna captured their attention. Those with a short attention span could concentrate better due to excitement in learning. This, too, is the result of incorporating nature into the learning programme.

In conclusion, learning with and in nature was helpful for this group of students. The learning program exhibited a therapeutic effect on these students. This shows that nature therapy can improve young people's learning ability.

12.7 Acknowledgements

Prof. Madya Dr. Norela Sulaiman of Head of UKM Research Centre Bukit Fraser for hosting this event. Team of Malaysia Genome Institute and UKM Research Centre Bukit Fraser scientist cum facilitators for practical hands-on activities and coordination. Andante Foundation and Khootz Outdoor Ventures as part of participants and facilitators.

12.8 References

Jalil, A (2014). Beescape for Meliponines. *Singapore: Partridge Publishing.*

Jalil. A., & Roubik(ed), D. (2016). Handbook of Meliponiculture. *Akademi Kelulut Malaysia*

Kuo, F.E and Andrea Faber Taylor, A.F., 2003. A Potential Natural Treatment for Attention-Deficit/Hyperactivity Disorder: Evidence From a National Study. *American Journal of Public Health.* Vol 94: 1580 -1586.

Schreibman, L., Dawson, G., Stahmer, A. C., Landa, R., Roges, S.J., Mcgee, G.G., Kasari, C., Ingersoll, B., Kaiser, A.P., Bruinsma, Y., McNerney, E., Wetherby, A., Halladay, A. 2015. Naturalistic Developmental Behavioral Interventions: Empirically Validated Treatments for Autism Spectrum Disorder. *Journal of Autism and Developmental Disorders* volume 45,2411–2428

Rasmussen, C. (2008). Catalogue of Indo-Malayan/Australasian stingless bees (Hymenoptera: Apidae: Meliponini). *Zootaxa 1935.*

Smith, D. R. (2012). Key to works of Indo-Malayan Stingless Bee – v.1.1.*11th International Conference of the Asian Apicultural Association. Kuala Terengganu*

Xiong, C. S. (2015). A Conservation Assessment of Stingless Bees (Apidae: Meliponini) in Singapore. *Cohort AY2015/2015 S1.*

Chapter 13

13.0 Diversity in Stingless Bee Honey Profiles and Implications of Trehalulose Discovery as One of The Main Sugars in Stingless Bee Honey.

Nurul Ainaa Farhanah Mat Ramlan[1], Norhasnida Zawawi,[2]*

[1]Department of Food Science, Faculty of Food Science and Technology, Universiti Putra Malaysia, 43400 Serdang, Selangor, Malaysia

[2]Natural Medicines and Product Research Laboratory, Institute of Bioscience, Universiti Putra Malaysia, 43400 Serdang, Selangor, Malaysia

*Corresponding author can be reached at norhasnida@upm.edu.my

13.1 Abstract

Stingless bee honey (SBH) is widely recognised as a wholesome food found to have many beneficial attributes, such as high antioxidant activities. Many researchers have previously reported sugar profiles of stingless bee honey from various species and locations with different nectar sources. Most of them said the common sugars found in the honey such as fructose, glucose, sucrose and maltose. However, the sugar composition in SBH is yet to be investigated thoroughly because of the recent discovery of trehalulose, a disaccharide sugar identified as one of the significant sugar compositions in SBH. Therefore, this chapter will compare sugar profiles for different stingless bees with other locations and nectars sources and introduce trehalulose discovered in SBH.

Keywords: Stingless bee, honey, sugar profile, trehalulose

13.2 Introduction

Stingless bee honey (SBH) is unique due to the plants' sources that stingless bees collect. They have their specific sweetness combine with an acidic taste (Abd Jalil, Kasmuri, & Hadi, 2017; Yaacob, Rajab, Shahar, & Sharif, 2018). This is because the stingless bee is smaller in size and could have the ability to pollinate a smaller-sized flower, which cannot be reached by honeybees except[68] for some Meliponini of genus *Melipona* that have more prominent size compared to some of *Apis* species (Bradbear, 2009; Vit, Pedro, & Roubik, 2013). The honey produced for stingless bee honey is slightly different from honeybees in colours, taste, and viscosity. They have a higher percentage in moisture content than *Apis* species that affects the characteristics of honey-like taste and consistency (Ávila et al., 2018; Nascimento et al., 2015). Codex Alimentarius Commission (2001) stated that the moisture

---

Editor's notes

[68] except when they are the same size as an *A. mellifera, koschevnikovi, nuluensis, cerana, indica, andreniformis* or *florea*, which around 90 species are...

content of *Apis* honey must not exceed 20%, and Malaysian Standard, MS2683: 2017 states that SBH must not exceed 35% of moisture content for raw and 22% for processing. It also has high acidity with a lower pH value, which provides the honey's ability to hinder the growth of microorganisms and contribute to the sourness and acidic taste of SBH (da Silva et al., 2013; Fatima et al., 2018).

Besides water, sugar is also one of the major compositions in SBH, with fructose and glucose as the common sugars present in high amounts. Despite containing fructose and glucose as the primary sugar, some studies have reported that the total of fructose and glucose in stingless bee honey is still lower than the requirement set by the international CODEX standard for honey. However, the Malaysian Standard for Kelulut (Stingless bee) honey, the MS2683:2017, has only put a maximum limit of not more than 85g/100g sucrose and fructose. Thus, the naturally lower total fructose and glucose content in some stingless bee kinds of honey can meet this standard's requirement. However, the lower total fructose and glucose content was not identified before discovering trehalulose (Fletcher et al., 2021).

Interestingly, to support the reason for the lower total fructose and glucose content, some studies had reported on maltose as one of the main sugar compositions in stingless bee honey (Oddo et al., 2008; Chuttong et al., 2016; Shamsudin et al., 2019; Wong et al., 2019). Recent findings have uncovered maltose sugar's confusion with new sugar called trehalulose (Fletcher et al., 2020). Therefore, this review helps explain each stingless bee species and the compilation of the sugar profile of stingless bee honey from different species, origin and nectar sources published before introducing the new founded sugar in stingless bee honey trehalulose.

13.3 Difference between stingless bee species that leads to the production of different honey quantities and qualities.

13.3.1 Species of stingless bees

Stingless bees (Meliponini) have extensive and diverse have the largest and diverse taxon consisting of 60 genera that are poorly known (Rasmussen and Cameron, 2010). Stingless bees can be found in tropical and sub-tropical regions of the world, such as in South and Central America, Africa, Southwest Asia and Australia (Ávila et al., 2018; Rasmussen and Cameron, 2010). According to Rasmussen (2008), 45 species of stingless bees from 14 genera have been recorded and identified in Malaysia. Kelly et al. (2014) stated five stingless bees, *Geniotrigona thoracica, Heterotrigona itama, Lepidotrigona terminata, Tetragonula fuscobalteata* and *Tetragonula leaviceps,* being commercialised in the meliponiculture industry for honey production and pollination in Malaysia.

13.4 Physical differences of stingless bees

13.4.1 Morphology of the Meliponini tribe.

According to Wille (1983), the stingless bee can be differentiated easily from other bees. They possess weak wing venation, the presence of the penicillium, long setae on the hind tibia, and reduction of the sting. Samsudin et al. (2017) have done a taxonomy study on the selected species of stingless bee in peninsular Malaysia. Some of the species are *Geniotrigona thoracica, Heterotrigona itama, Lepidotrigona terminata* and *Tetragonula laeviceps.*

Based on their finding, *Geniotrigona thoracica* has a relatively large size compared to other stingless bee species, 8.44 mm. The presence of two vertical black stripes at the middle of the entirely brown mesoscutum is one way to identify *G. thoracica.* It has a very long forewing with 8.22 mm and an entirely black hind tibia with and 3.26 mm. The average body size of *H. itama* is 5.58 mm. It has blackish-brown colouration on the tarsal tergite and white colouration of hair on frons. For *Tetragonula laeviceps*, they are characterised as the small stingless bee with an average of 4.31 mm. It possessed only one distinct vertical band in the mesoscutum, and the first to sixth gastral tergite is completely dark brown. *Lepidotrigona terminata* has an average of 6.29 mm of body length. It has a black scutellum entirely, which framing with a golden yellow scale-like at the posterior and black colour of frons and clypeus covered with thick white hairs (Samsudin et al., 2017).

13.5 Physicochemical properties of some stingless bee honey: Comparison across species

13.5.1 Comparison of stingless bee honey composition based on the species, origin and nectar preferences

Stingless bee honey is reported to have a major fructose and glucose composition with other minor constituents like phenolic compounds, enzymes, organic acids, minerals, and pollens (Ávila et al., 2018; de Sousa et al., 2016). Based on table 1, fructose and glucose sugar are mainly found in stingless bee honey[69]. The total fructose and glucose for stingless bee honey from Brazil, Venezuela and Thailand are high, where most of them have amounts ranging from 40% to 70% (w/w). In contrast, fructose and glucose from Malaysia, Indonesia, and Australia have low amounts of less than 60% (w/w).

---

[69]Editor's notes

[69] There are a few samples, there are over 600 species and the flora used during different times of the year differs, so that there are 1000s of different 'kinds' of stingless bee honeys. Your wording gives the impression that we already know stingless bee honey. As everyone who contributed to the first volume dedicated to the subject (Vit, Pedro and Roubik, 2013) knows, we are just beginning to form an opinion on the total content of meliponine honeys, and their origins.

Sucrose sugar is present in most stingless bee honey, but it has a low amount except for stingless bee honey reported by Cheng et al. (2019). They reported that stingless bee honey (*Heterotrigona itama*) from Muar, Malaysia has the highest sucrose amount with 25.63 g/100g. This sucrose amount exceeds the limit stated in Malaysian Standard (MS2683:2017), which sucrose for raw stingless bee honey must not exceed 7.5g/100g, and processed honey must not exceed 8.0 g/100g. Sucrose content in honey can indicate honey maturity because high sucrose in honey can indicate that the honey is not yet mature (Almeida-Muradian, 2013; Dardón, Maldonado-Aguilera, and Enríquez, 2013).

There are various sources of nectar from plants that stingless bee honey collected and transformed into honey. This factor might contribute to the different proportion of sugars in stingless bee honey and other components such as phenolic and flavonoid compounds (da Silva et al., 2013; Wong et al., 2019). Acacia[70] (*Acacia mangium*) is a flowering tree that becomes one of the most pollen sources preferred by Malaysia's stingless bee. Majid et al. (2020) found out that acacia has one of the predominant pollens chosen by *Heterotrigona itama*. Other nectars involved are fruit trees such as starfruit (*Averrhoa carambola L*), coconut (*Cocos nucifera*) and African oil palm (*Elaeis guineensis*), as well as shrubs or small tree-like mangrove (*Rhizophora mucronata*) and famous flowering plant, Mexican creeper (*Antigonon leptopus*) (Majid et al., 2020; Majid et al., 2019; Shamsudin et al., 2019).

The sugar composition of the stingless bee honey is different for each species as each species' geographical region is different. As the origin of each species is different, the nectar sources also vary. Other factors such as climate and harvesting season affect the stingless bee honey's physical and chemical composition (Ball, 2007; Kaškoniene and Venskutonis, 2010).

13.5.2 Discovery of trehalulose sugar.

The first discovery of trehalulose was by Lund and Wyatt (1973) when they grew a strain of *Erwinia carotovora* subspecies *atroseptica* in media with high sucrose concentration and found out that it produces both isomaltulose (IM) and trehalulose sugar (Byrne, Hendrix, and Williams, 2003). The study by Fuji et al. (1983) on the oligosaccharides formed by *Serratia plymuthica* cited in Cheetham (1984) revealed that IM is the major product, and trehalulose, isomaltose, isomelezitose are the minor products in the process. This shows that trehalulose was produced as the secondary product

Editor's notes

[70] Unfortunately, *Acacia* has no floral nectar. *Robinia* has nectar, and it used to be placed in the genus *Acacia*, but *A. mangium* has extrafloral nectar, only. Its flowers have pollen. Please correct this notion if it is incorrect, but it seems the idea of Acacia nectar persists, and it needs to be corrected. If pollen of a plant that lacks floral nectar (such as *Mimosa, Acacia, Solanum, Cassia*, many legumes, palms and Melastomataceae) shows up in honey, that means absolutely nothing, as far as a nectar source is concerned.

in the process of enzymic transglucosylation of sucrose to IM (Nakajima et al., 1990). Trehalulose also naturally exists in small quantities in honey (Nakajima et al., 1990; Wei et al., 2013) and as a major component in *Bemisia* honeydew (Salvucci, Wolfe, and Hendrix, 1997).

Recently, trehalulose sugar has been identified as the major sugar component in stingless bee honey, comprising between 13 and 44g per 100g honey (Fletcher et al., 2020). They also reported that stingless bee honey from Malaysia (*Geniotrigona thoracica*) has the highest composition of trehalulose with 84% (w/w) of the total sugar content out of the four kinds of honey produced by different species that were studied. Trehalulose or 1-O-α-D-glucopyranosyl-D-fructose also a reducing disaccharide that has two tautomers, which are 1-O-α-D-glucopyranosyl-β-D-fructopyranose as major form (about 66%) and 1-O-α-D-glucopyranosyl-β-D-fructofuranose as minor form (about 33%) (Cookson, Cheetham, and Rathbone, 1987). Trehalulose has been reported to have around 60% sweetness compared to sucrose (Tian et al., 2019; Wei et al., 2013).

13.5.3 Inaccurate reports of maltose in stingless bee honey

A recent finding by Fletcher et al. (2020) reported that they had identified a single disaccharide that is present in each honey they had tested. The disaccharide did not match any of their available disaccharide standards. According to the previous study done by Oddo et al. (2008) on *Tetragonula carbonaria*, honey reported that 20% of sugar than glucose and fructose is considered maltose even though the retention time does not match with the maltose standard. Earlier research by Bogdanov, Vit, & Kilchenmann (1996) found that stingless bee honey has predominant sugars of fructose and glucose with traces of disaccharides *Trigona* and most non-*meliponine* have a different spectrum with main sugar identified is a disaccharide stated as maltose. Vit, Fernandez-Maeso, & Ortiz-Valbuena (1998) also found that the *Trigonini* kinds of honey have a high amount of disaccharide maltose compared to *Meliponini* kinds of honey.

Nevertheless, the main disaccharides identified by Fletcher et al. (2020) using UPLC-MS/MS has successfully excluded maltose. They characterised the disaccharides present in the SBH using Nuclear Magnetic Resonance (NMR) spectroscopy as trehalulose. Therefore, this finding had cleared the confusion that occurred previously on maltose as the major sugar or major disaccharide present in SBH as reported in findings by Chuttong et al. (2016); Se et al. (2018); Shamsudin et al. (2019) and Wong et al., (2019).

13.6 Tabulations

Table 1. Sugar profile of stingless bee honey from different species, source of nectars and origin.

Species	Source of nectar/ predominant nectar/pollen	Sample location	Sugar (g/100g)								Origin	Reference
			Fructose	Glucose	Sucrose	Maltose	Arabinose	Turanose	Trehalose	Erlose		
Heterotrigona itama	Not mentioned	S1=Kuala Nerang	16.03 ± 1.64	18.89 ± 2.25	6.72 ± 2.20	ND	ND	ND	ND	ND	Malaysia	(Cheng et al., 2019)
		S2=UPM	28.91 ± 1.58	17.06 ± 1.85	9.30 ± 0.93	ND	ND	ND	ND	ND		
		S3=Muar	33.13 ± 1.76	20.36 ± 2.16	25.63 ± 1.23	ND	ND	ND	ND	ND		
		S4= Kuala Terengganu	16.19 ± 3.08	8.63 ± 2.30	6.50 ± 0.56	ND	ND	ND	ND	ND		
		F1= Tumpat	25.91 ± 1.49	16.51 ± 1.63	7.77 ± 1.99	ND	ND	ND	ND	ND		
		F2= Sibu	27.76 ± 2.13	17.01 ± 2.02	6.91 ± 1.80	ND	ND	ND	ND	ND		
		F3= Kota Belud	26.20 ± 0.64	16.18 ± 4.20	2.68 ± 0.73	ND	ND	ND	ND	ND		
Geniotrigona thoracica	Acacia (*Acacia mangium*)	Terengganu	7.89 ± 0.36	7.89± 0.36	ND	28.96 ± 1.17	ND	ND	ND	ND	Malaysia	(Shamsudin et al., 2019)

	Starfruit (*Averrhoa carambola L*)	Pahang	9.34 ± 0.06	9.34 ± 0.06	ND	40.82 ± 0.79	ND	ND	ND	ND		
	Gelam (*Melaleuca cajaputi*)	Johor	8.43 ± 0.11	8.43 ± 0.11	ND	43.15 ± 2.36	ND	ND	ND	ND		
Heterotrigona itama	Acacia (*Acacia mangium*)	Terengganu	13.00 ± 0.21	13.00 ± 0.21	ND	22.56 ± 1.22	ND	ND	ND	ND		
	Starfruit (*Averrhoa carambola L*)	Pahang	9.68 ± 0.72	9.68 ± 0.72	ND	37.35 ± 0.58	ND	ND	ND	ND		
	Gelam (*Melaleuca cajaputi*)	Johor	13.58 ± 0.33	13.58 ± 0.33	ND	31.50 ± 2.11	ND	ND	ND	ND		
Heterotrigona itama	*Acacia mangium*	Nanga Dap, Kanowit (n=45), Tanjung Manis (n=51)	22.00 ± 3.48	23.44 ± 3.23	0.04 ± 024	27.41 ± 8.32	ND	ND	ND	ND	Malaysia	(Wong et al., 2019)
Tetrigona binghami	*Acacia mangium*	Nanga Dap, Kanowit (n=5), Tanjung Manis (n=5)	15.43 ± 3.97	19.84 ± 4.87	ND	27.95 ± 3.47	ND	ND	ND	ND		

Homotrigona fimbriata	Not mentioned	Chiang Mai (n=1)	7.4	15	ND	ND	ND	ND	ND	ND	Thailand	(Chuttong et al., 2016)
Lepidotrigona terminata		Chanthaburi (n=1)	8.1	4.9	ND	53	ND	ND	ND	ND		
Lepidotrigona flavibasis		Chiang Mai, Chanthaburi (n=4)	16±8.8	13±8.6	ND	39±15	ND	ND	ND	ND		
Lepidotrigona doipaensis		Chiang Mai (n=2)	15,9	14,9.8	ND	ND	ND	ND	ND	ND		
Lisotrigona furva		Chanthaburi (n=2)	34,33	25,28	ND, 6.0	ND	ND	ND	ND	ND		
Tetragonilla collina		Chiang Mai (n=1)	26	26	ND	ND	ND	ND	ND	ND		
Tetragonula fuscobalteata		Chanthaburi (n=2)	23,19	19,4.4	ND	ND	ND	ND	ND	ND		
Tetragonula leaviceps-pagdeni complex		Chiang Mai, Chanthaburi, Trat (n=10)	17±67	12±7.7	ND, 0.025	37±12	ND	ND	ND	ND		
Tetragonula testaceitarsis		Chanthaburi (n=2)	15,29	11,27	ND	ND	ND	ND	ND	ND		
Tetrigona apicalis		Chiang Mai (n=2)	6.8,6.7	7.7,4.1	ND	ND	ND	ND	ND	ND		
Tetrigona melanoleuca		Chiang Mai (n=1)	6	8.9	ND	ND	ND	ND	ND	ND		

Melipona subnitida	*Ziziphus joazeiro* Mart (juazeiro)	semiarid region of Brazilian northeastern (Serido region, state of Rio Grande do Norte and Agreste region of state of Paraíba)	59.2 ± 1.1	37.7 ± 0.4	1.6 ± 0.2	ND	ND	ND	ND	ND	Brazil	(de Sousa et al., 2016)
	Mimosa quadrivalvis L. (malícia)		50.0 ± 0.9	45.4 ± 0.3	3.9 ± 0.2	ND	0.5 ± 0.0	ND	ND	ND		
	Croton heliotropiifolius Kunth (velame branco)		55.7 ± 1.9	42.1 ± 0.6	0.7 ± 0.0	ND	1.0 ± 0.1	ND	ND	ND		
	Mimosa arenosa (willd.) Poir (jurema branca)		52.6 ± 2.5	45.7 ± 0.4	1.2 ± 0.2	ND	0.4 ± 0.1	ND	ND	ND		
Melipona scutellaris	*Ziziphus joazeiro* Mart (juazeiro)		57.6 ± 1.5	38.1 ± 0.3	2.6 ± 0.5	ND	0.3 ± 0.0	ND	ND	ND		
	Mimosa quadrivalvis L. (malícia)		55.5 ± 1.0	42.6 ± 0.2	1.9 ± 0.3	ND	0.4 ± 0.1	ND	ND	ND		
	Croton heliotropiifolius Kunth		53.8 ± 0.9	43.3 ± 0.1	2.0 ± 0.3	ND	0.7 ± 0.1	ND	ND	ND		

	(velame branco)											
	Mimosa arenosa (willd.) Poir (jurema branca)		53.6 ± 0.7	41.4 ± 0.5	3.0 ± 0.5	ND	0.2 ± 0.0	ND	ND	ND		
Melipona bicolar	Sylvan	Florianópolis	32.5 ± 0.220	29.5 ± 0.160	<LOQ	ND	ND	ND	ND	ND	Brazil	(Biluca, Braghini, Gonzaga, Costa, & Fett, 2016b)
	Sylvan	São Miguel do Oeste	31.5 ± 0.200	29.5 ± 0.160	<LOQ	ND	ND	ND	ND	ND		
	Eucalyptus spp. and *Mimosa bimucronata*	Santa Rosa de Lima	33.2 ± 0.350	27.3 ± 0.770	<LOQ	ND	ND	ND	ND	ND		
	Eucalyptus spp.		34.7 ± 0.740	26.6 ± 0.590	<LOQ	ND	ND	ND	ND	ND		
	Eucalyptus spp.		32.3 ± 1.00	26.5 ± 0.850	<LOQ	ND	ND	ND	ND	ND		
Scaptotrigona bicunctata	Sylvan	Florianópolis	31.1 ± 0.240	27.1 ± 0.130	<LOQ	ND	ND	ND	ND	ND		
	Sylvan	São Miguel do Oeste	40.2 ± 0.980	27.3 ± 0.100	<LOQ	ND	ND	ND	ND	ND		
Melipona quadriasciata	Sylvan	Florianópolis	35.1 ± 0.800	27.3 ± 0.090	<LOQ	ND	ND	ND	ND	ND		

	Sylvan	Santo Amaro da Imperatriz	38.4 ± 0.120	31.1 ± 0.060	<LOQ	ND	ND	ND	ND	ND		
	Sylvan	São Miguel do Oeste	31.9 ± 0.190	26.5 ± 0.170	<LOQ	ND	ND	ND	ND	ND		
	Citrus sinensis	Santa Rosa de Lima	32.7 ± 1.11	27.9 ± 0.920	<LOQ	ND	ND	ND	ND	ND		
	Hovena dulcis		35.1 ±0.470	31.3 ± 0.630	<LOQ	ND	ND	ND	ND	ND		
	Eucalyptus spp.		30.4 ± 0.200	25.6 ± 0.530	<LOQ	ND	ND	ND	ND	ND		
	Mimosa bimucronata		30.7 ±0.170	24.1 ± 0.410	<LOQ	ND	ND	ND	ND	ND		
	Citrussinensis		37.3 ± 0.270	29.2 ± 0.160	<LOQ	ND	ND	ND	ND	ND		
	Hovena dulcis and *Vernonanthua discolor*		39.4 ± 0.560	30.2 ± 0.240	<LOQ	ND	ND	ND	ND	ND		
	Sylvan		35.2 ±0.680	24.6 ± 0.290	<LOQ	ND	ND	ND	ND	ND		
	Sylvan		37.1 ± 0.620	25.4 ± 0.240	<LOQ	ND	ND	ND	ND	ND		
	Hovena dulcis and		33.2 ±0.590	25.8 ± 0.890	<LOQ	ND	ND	ND	ND	ND		

	Vernonanthua discolor											

Melipona monduri	Sylvan	Florianópolis	36.4 ±0.150	29.5 ± 0.130	<LOQ	ND	ND	ND	ND	ND	Brazil	(Biluca et al., 2016b)
	Eucalyptus spp.	Santa Rosa de Lima	38.9 ± 0.300	30.1 ± 0.400	<LOQ	ND	ND	ND	ND	ND		
Melipona rufiventris mondory	Sylvan	São Miguel do Oeste	35.6 ±0.120	30.1 ± 0.080	<LOQ	ND	ND	ND	ND	ND		
Tetragona claviceps	Sylvan	São Miguel do Oeste	40.2 ± 0.640	8.21 ± 0.200	<LOQ	ND	ND	ND	ND	ND		
Melipona scutellaris	Sylvan	São Miguel do Oeste	36.38 ± 0.360	26.3 ± 0.280	<LOQ	ND	ND	ND	ND	ND		
Trigona fuscipennis	Sylvan	São Miguel do Oeste	36.22 ± 0.710	20.4 ± 0.970	<LOQ	ND	ND	ND	ND	ND		

Tetragonula laeviceps	Banana (*Musa paradisiaca L.*) Rambutan (*Nephelium lappaceum*) Canarium (*Canarium indicum L.*) Tamarind (*Tamarindus indica*) Matoa (*Pometia pinnata*) Cattapa (*Terminalia catappa*) Caimito (*Chrysophyllum cainito*)	Faculty of Animal Science UGM	22.92	20.25	ND	ND	ND	ND	ND	ND	Indonesia	(Agussalim, Agus, Nurliyani, & Umami, 2019)

	Coconut (*Cocos nucifera*) Mango (*Mangifera indica L.*) Kapok (*Ceiba pentandra*) Cashew (*Anacardium occidentale*)	Lombok, West Nusa Tenggara	20.76	11.49	ND	ND	ND	ND	ND	ND		
	Calliandra (*Calliandra calothyrsus*) Mexican creeper (*Antigonon leptopus*) Banana (*Musa paradisiaca L.*)	Nglipar, Yogyakarta	7.79	22.78	ND	ND	ND	ND	ND	ND		

	Mango (*Mangifera indica L.*) White albizia (*Paraserianthes falcataria L.*)											
Melipona compressipes (n=5)	Not mentioned	different regions of Venezuela	36.6 ± 1.0	36.3 ± 2.8	0.1	0.1	ND	3.2 ± 0.6	-	0.1 ± 0.1	Venezuala	(Bogdanov, Vit, & Kilchenmann, 1996)
Melipona trinitatis (n=5)			36.8 ± 1.8	36.9 ± 2.1	-	0.2 ± 0.1	ND	2.5 ± 1.6	-	-		
Melipona favosa (n=14)			34.8 ± 2.5	36.9 ± 2.7	0.1	0.1	ND	2.8 ± 1.0	-	-		
Frieseomelitta aff. *varia* (n=8)			24.4± 4.1	18.1 ± 6.9	0.2	0.4	ND	32.3 ± 11.4	0.2 ± 0.2	0.5 ± 0.3		
non-Melipona (mostly Trigona) (n=10)			30.4 ± 8.0	26.4 ± 8.6	0.1	0.4	ND	19.7 ± 16.0	0.1	0.1		
Melipona compressipes manaosensis	Not mentioned	Manacapuru and Itacoatiara	32.04 ± 0.25	28.59 ± 0.50	0.21 ± 0.03	ND	ND	ND	ND	ND	Brazil	(Ligia B. De Almeida-

Melipona seminigra merrillae		villages, in Amazonas state, north Brazil	30.72 ± 0.20	29.44 ± 0.41	0.08 ± 0.03	ND	ND	ND	ND	ND		Muradian, Matsuda, & Bastos, 2007)
Heterotrigona itama	Not mentioned	Johor (n=43) Terengganu (n=14) Kedah (n=3) Pahang (n=1) Kelantan (n=1)	13.4 ± 3.83	16.4 ± 5.86	3.46 ± 2.49	33.8 ± 10.6	ND	ND	ND	ND	Malaysia	(Se et al., 2018)
Tetragonula carbonaria	mainly ornamental shrubs and flowering trees	various locations around Brisbane, Queensland, Australia (suburban area) (n=8)	24.5 ± 1.9	17.5 ± 2.8	1.8 ± 0.4	20.3 ± 2.9	ND	ND	ND	ND	Australia	(Oddo et al., 2008)
	Not mentioned (raw honey)	Not mentioned	12.39 ± 0.31	3.41 ± 0.06	2.37 ± 0.06	ND	ND	ND	ND	ND	Malaysia	(Syariffud deen Mhd

Stingless bee (not mentioned which species)	Not mentioned (honey after dehydrated)		13.52 ± 0.75	3.46 ± 0.21	2.48 ± 0.13	ND	ND	ND	ND	ND		Adnan et al., 2018)

13.7 Conclusion

Honey produced by stingless bees has very diverse sugar profiles, possibly because of the diversity in species type, geographical regions, flowering seasons, climate and harvesting seasons. The impact of these factors on individual sugar production in honey should be further investigated to understand better why sugar profiles in stingless bee honey are different from honeybee kinds of honey. As trehalulose sugar has been recently reported, future studies should also identify and quantify trehalulose as a possible main sugar in stingless bee honey produced by all species.

13.8 References

Abd Jalil, M. A., Kasmuri, A. R., & Hadi, H. (2017). Stingless bee honey, the natural wound healer: A review. *Skin Pharmacology and Physiology*, *30*(2), 66–75. https://doi.org/10.1159/000458416

Agussalim, A., Agus, A., Nurliyani, N., & Umami, N. (2019). The sugar content profile of honey produced by the Indonesian Stingless bee, Tetragonula laeviceps, from different regions. *Livestock Research for Rural Development*, *31*(6), 1–6.

Almeida-Muradian, Ligia Bicudo de. (2013). Chapter 26: Tetragonisca angustula Pot-Honey Compared to Apis mellifera Honey from Brazil. In Vit, Patricia, Pedro, Silvia. R. M., & Roubik, David. W, *Pot-Honey: A legacy of stingless bees* (1st, ed., pp. 375–382). © Springer-Verlag New York.

Ávila, S., Beux, M. R., Ribani, R. H., & Zambiazi, R. C. (2018). Stingless bee honey: Quality parameters, bioactive compounds, health-promotion properties and modification detection strategies. *Trends in Food Science and Technology*, *81*(March), 37–50. https://doi.org/10.1016/j.tifs.2018.09.002

Ball, D. W. (2007). The chemical composition of honey. *Chemical Reviews*, *84*(10), 1643–1646. https://doi.org/10.1021/cr60125a003

Biluca, F. C., Braghini, F., Gonzaga, L. V., Costa, A. C. O., & Fett, R. (2016a). Physicochemical profiles, minerals and bioactive compounds of stingless bee honey (Meliponinae). *Journal of Food Composition and Analysis*, *50*, 61–69. https://doi.org/10.1016/j.jfca.2016.05.007

Biluca, F. C., Braghini, F., Gonzaga, L. V., Costa, A. C. O., & Fett, R. (2016b). Physicochemical profiles, minerals and bioactive compounds of stingless bee honey (Meliponinae). *Journal of Food Composition and Analysis*, *50*, 61–69. https://doi.org/10.1016/j.jfca.2016.05.007

Bogdanov, S., Vit, P., & Kilchenmann, V. (1996). Sugar profiles and conductivity of stingless bee honey from Venezuela. *Apidologie*, *27*(6), 445–450. https://doi.org/10.1051/apido:19960602

Bradbear, N. (2009). Bees and their role in forest livelihoods. Food and Agriculture Organization of the United Nations.

Byrne, D. N., Hendrix, D. L., & Williams, L. H. (2003). Presence of trehalulose and other oligosaccharides in hemipteran honeydew, particularly Aleyrodidae. *Physiological Entomology*, *28*(2), 144–149. https://doi.org/10.1046/j.1365-3032.2003.00310.x

Cheng, M. Z. S. Z., Ismail, M., Wei Chan, K., Jiun Ooi, D., Ismail, N., Zawawi, N., & Mohd Esa, N. (2019). Comparison of Sugar Content, Mineral Elements and Antioxidant Properties of Heterotrigona Itama Honey from Suburban and Forest in Malaysia. *Malaysian Journal of Medicine and Health Sciences*, *15*(SP1), 2636–9346.

Cheetham, P. S. J. (1984). The extraction and mechanism of a novel isomaltulose-synthesising enzyme from Erwinia rhapontici. *The Biochemical Journal*, *220*(1), 213–220. https://doi.org/10.1042/bj2200213

Chuttong, B., Chanbang, Y., Sringarm, K., & Burgett, M. (2016). Physicochemical profiles of stingless bee (Apidae: Meliponini) honey from South East Asia (Thailand). *Food Chemistry*, *192*, 149–155. https://doi.org/10.1016/j.foodchem.2015.06.089

Codex Alimentarius Commission. (2001). Codex Standard for Honey, CODEX STAN 12-1981. *Codex Alimentarius Commission FAO/OMS*, 1–8.

Cookson, D., Cheetham, P. S. J., & Rathbone, E. B. (1987). Preparative high-performance liquid chromatographic purification and structural determination of 1-O-α-d-glucopyranosyl-d-fructose (trehalulose). *Journal of Chromatography*, *402*, 265–272. https://doi.org/10.1016/S0167-2991(09)60435-3

da Silva, I. A. A., da Silva, T. M. S., Camara, C. A., Queiroz, N., Magnani, M., de Novais, J. S., … de Souza, A. G. (2013). Phenolic profile, antioxidant activity and palynological analysis of stingless bee honey from Amazonas, Northern Brazil. *Food Chemistry*, *141*(4), 3552–3558. https://doi.org/10.1016/j.foodchem.2013.06.072

Dardón, M. J., Maldonado-Aguilera, C., & Enríquez, E. (2013). Chapter 28 : The Pot-Honey of Guatemalan Bees. In *Pot-Honey: A Legacy of Stingless Bees* (pp. 395–408).

De Almeida-Muradian, Ligia B., Matsuda, A. H., & Bastos, D. H. M. (2007). Physicochemical

Parameters of Amazon Melipona Honey. *Quim Nova*, *30*(3), 707–708.

De Sousa, J. M. B., de Souza, E. L., Marques, G., Benassi, M. de T., Gullón, B., Pintado, M. M., & Magnani, M. (2016). Sugar profile, physicochemical and sensory aspects of monofloral honey produced by different stingless bee species in the Brazilian semi-arid region. *LWT - Food Science and Technology*, *65*, 645–651. https://doi.org/10.1016/j.lwt.2015.08.058

Fatima, I. J., Mohd Hilmi, A. B., Salwani, I., & Lavaniya, M. (2018). Physicochemical characteristics of Malaysian stingless bee honey from Trigona species. *International Medical Journal Malaysia*, *17*, 187–191.

Fletcher, M. T., Hungerford, N. L., Webber, D., Jesus, M. C. De, Zhang, J., Stone, I. S. J., … Zawawi, N. (2020). Stingless bee honey, a novel source of trehalulose : a biologically active disaccharide with health benefits. *Scientific Reports*, *10*(1), 1–8. https://doi.org/10.1038/s41598-020-68940-0

Kaškoniene, V., & Venskutonis, P. R. (2010). Floral Markers in Honey of Various Botanical and Geographic Origins: A Review. *Comprehensive Reviews in Food Science and Food Safety*, *9*(6), 620–634. https://doi.org/10.1111/j.1541-4337.2010.00130.x

Kelly, N., Farisya, M. S. N., Kumara, T. K., & Marcela, P. (2014). Species diversity and external nest characteristics of stingless bees in meliponiculture. *Pertanika Journal of Tropical Agricultural Science*, *37*(3), 293–298.

Lund, B. M., & Wyatt, G. M. (1973). The nature of reducing compounds formed from sucrose by Erwinia carotovora var. atroseptica. *Journal of General Microbiology*, *78*(2), 331–336. https://doi.org/10.1099/00221287-78-2-331

Majid, M., Ellulu, M. S., & Abu Bakar, M. F. (2020). Melissopalynological Study, Phenolic Compounds, and Antioxidant Properties of Heterotrigona itama Honey from Johor, Malaysia. *Scientifica*, *2020*, 1–9. https://doi.org/10.1155/2020/2529592

Majid, M., Fadzelly Abu Bakar, M., Mian, Z., Esa, F., & Kok Yeow, Y. (2019). Variations of physicochemical properties of stingless bee honey from a different botanical origin in Johor, Malaysia. *IOP Conference Series: Earth and Environmental Science*, *269*(1). https://doi.org/10.1088/1755-1315/269/1/012028

Nakajima, Y., Sugitani, T., Tanaka, M., & Fujii, S. (1990). The occurrence of Trehalulose, 1-0-α-D-Glucopyranosyl D-Fructose, in Nectar Honey. *Shokuhin Kogyo Gakkaishi*, *37*(7), 354–358.

Nascimento, A., Marchini, L., Carvalho, C., Araújo, D., Olinda, R., & Silveira, T. (2015). Physical-

Chemical Parameters of Honey of Stingless Bee (Hymenoptera: Apidae). *American Chemical Science Journal*, *7*(3), 139–149. https://doi.org/10.9734/acsj/2015/17547

Oddo, L. P., Heard, T. A., Rodríguez-Malaver, A., Pérez, R. A., Fernández-Muiño, M., Sancho, M. T., … Vit, P. (2008). Composition and antioxidant activity of Trigona carbonaria honey from Australia. *Journal of Medicinal Food*, *11*(4), 789–794. https://doi.org/10.1089/jmf.2007.0724

Omar, S., Enchang, F. K., Nazri, M. U. I. A., Ismail, M. M., & Ismail, W. I. W. (2019). Physicochemical profiles of honey harvested from four major species of stingless bee (Kelulut) in northeast peninsular of Malaysia. *Malaysian Applied Biology*, *48*(1), 111–116.

Rasmussen, C. (2008). *Catalogue of the Indo-Malayan/Australasian stingless bees (Hymenoptera: Apidae: Meliponini)*. *Zootaxa*. Magnolia Press. https://doi.org/10.11646/zootaxa.1935.1.1

Rasmussen, C., & Cameron, S. A. (2010). Global stingless bee phylogeny supports ancient divergence, vicariance, and long-distance dispersal. *Biological Journal of the Linnean Society*, *99*, 206–232. https://doi.org/10.1111/j.1095-8312.2009.01341.x

Salvucci, M. E., Wolfe, G. R., & Hendrix, D. L. (1997). Effect of sucrose concentration on carbohydrate metabolism in Bemisia argentifolii: Biochemical mechanism and physiological role for trehalulose synthesis in the Silverleaf whitefly. *Journal of Insect Physiology*, *43*(5), 457–464. https://doi.org/10.1016/S0022-1910(96)00124-2

Samsudin, S. F., Mamat, M. R., & Hazmi, I. R. (2017). Taxonomic Study on Selected Species of Stingless Bee (Hymenoptera: Apidae: Meliponini) In Peninsular Malaysia. *Serangga*, *23*(2), 203–258.

Se, K. W., Ibrahim, R. K. R., Wahab, R. A., & Ghoshal, S. K. (2018). Accurate evaluation of sugar contents in stingless bee (Heterotrigona itama) honey using a swift scheme. *Journal of Food Composition and Analysis*, *66*, 46–54. https://doi.org/10.1016/j.jfca.2017.12.002

Shamsudin, S., Selamat, J., Sanny, M., Abd. Razak, S. B., Jambari, N. N., Mian, Z., & Khatib, A. (2019). Influence of origins and bee species on physicochemical, antioxidant properties and botanical discrimination of stingless bee honey. *International Journal of Food Properties*, *22*(1), 238–263. https://doi.org/10.1080/10942912.2019.1576730

Syariffuddeen Mhd Adnan, A., Hamzah, A., Sahari, Y., Ismail, N., Hafifi Maidin, K., Jajuli, R., … Aliq Jamaluddin dan Shafie Alwi, M. (2018). Alat penyahidrat sistematik untuk madu kelulut (Systematic dehydrator for stingless bee honey). *Buletin Teknologi MARDI*, *13*, 1–11. Retrieved from http://ebuletin.mardi.gov.my/buletin/13/Alat penyahidrat.pdf

Tian, Y., Deng, Y., Zhang, W., & Mu, W. (2019). Sucrose isomers as alternative sweeteners: properties, production, and applications. *Applied Microbiology and Biotechnology*, *103*(21–22), 8677–8687. https://doi.org/10.1007/s00253-019-10132-6

Vit, P., Fernandez-Maeso, M. C., & Ortiz-Valbuena, A. (1998). Potential use of the three frequently occurring sugars in honey to predict stingless bee entomological origin. *Journal of Applied Entomology*, *122*(1), 5–8. https://doi.org/10.1111/j.1439-0418.1998.tb01452.x

Vit, Patricia, Pedro, Silvia. R. M. & Roubik, David. W. (2013). *Pot-Honey: A Legacy of Stingless Bees* (1st ed.). Springer-Verlag New York.

Wei, Y., Liang, J., Huang, Y., Lei, P., Du, L., & Huang, R. (2013). Simple, fast, and efficient process for producing and purifying trehalulose. *Food Chemistry*, *138*(2–3), 1183–1188. https://doi.org/10.1016/j.foodchem.2012.11.115

Wille, A. (1983). Biology of the Stingless Bees. *Annual Review of Entomology*, *28*, 41–64. https://doi.org/10.1146/annurev.en.28.010183.000353

Wong, P., Ling, H. S., Chung, K. C., Yau, T. M. S., & Gindi, S. R. A. (2019). Chemical Analysis on the Honey of Heterotrigona itama and Tetrigona binghami from Sarawak, Malaysia. *Sains Malaysiana*, *48*(8), 1635–1642. https://doi.org/10.17576/jsm-2019-4808-09

Yaacob, M., Rajab, N. F., Shahar, S., & Sharif, R. (2018). Stingless bee honey and its potential value: A systematic review. *Food Research*, *2*(2), 124–133. https://doi.org/10.26656/fr.2017.2(2).212

Chapter 14

14.0 Stingless Beekeeping Practices and Honey Industry From A Malaysian Perspective

Wan Iryani Wan Ismail[1,2], Muhammad Faiz Zulkifli[1,2] and Mohd Mansor Ismail[3]

[1]Cell Signalling and Biotechnology Research Group (CeSBTech), Faculty of Science and Marine Environment, Universiti Malaysia Terengganu, 21030 Kuala Nerus, Terengganu, Malaysia

[2]Biological Security and Sustainability (BioSeS) Research Group, Faculty of Science and Marine Environment, Universiti Malaysia Terengganu, 21030 Kuala Nerus, Terengganu, Malaysia

[3]Department of Agribusiness and Bioresource Economics, Faculty of Agriculture, Universiti Putra Malaysia, 43400 Serdang, Selangor, Malaysia

14.1 Introduction

The honey industry contributes significantly to the well-being of rural dwellers in Malaysia. It started with wild giant honeybees (*Apis dorsata*). Native honeybees (*A. cerana* and *A. florea*), followed by the imported European honeybees (*A. mellifera*) and now with the local stingless bees (*Meliponini*). More importantly, the industry aims to alleviate poverty among the rural poor. Therefore, strategic plans to sustain the industry and be beneficial to the targeted society must be executed well for the short- and long-term effects.

The honey industry started by the local bee hunters collecting honey from the jungle since the Malacca Sultanate in the 16th century based on the old literature. The practice is believed to start far earlier; however, no documented evidence in supporting the claim. Jungle honey harvested from *A. dorsata* is often used in Malay traditional medicine to treat various illnesses and was sold as one of the barter system's products. After a few developments, from honey hunting to modern beekeeping using domesticated honeybees and stingless bees, Malaysia's beekeeping practices have grown as one of the most dynamic industries, particularly in providing stingless bee honey or locally known as kelulut honey. The Department of Agriculture Malaysia reported about RM 20 million value of stingless bee honey was sold in 2017. Even though the amount is small, the actual figure is believed far higher than the reported amount because the data is only based on the shared data by the beekeepers to the department on a volunteer basis. Therefore, by maximizing data collection, and domestic resources for honey production and other by-products, the industry aims to be one of the new Malaysian commodities after rubber and oil palm in the coming future.

14.2 History and Beekeeping Practices in Malaysia

Beekeeping practices in Malaysia can be classified as traditional and modern. Traditional beekeeping involves bee hunting activity, where local bee hunters collect wild honey produced by honeybee species *A. dorsata* (giant wild bees) from the jungle at a specific time. For instance, in Kedah, the activity is conducted from December to April every year in the tropical forest. In Terengganu, more honey is available between May to December from Gelam (*Melaleuca cajaputi*) forest. The practice has been applied since ancient times until today, with slight differences probably due to climate change. The major difference is the time of harvest and the apparatus used to collect the honey. In Kedah, the traditional practice is still maintained until today, such as harvesting only during a moonless night, the use of cow-skin container and knife made from cow's bone. The honey hunter usually avoids applying any metal in the honey collection procedure as they believe that metal can reduce the medicinal effects of the collected honey.

In modern beekeeping, honey is harvested from reared bees (mostly honeybee species *A. mellifera*) in hives. Initially, a systematic local bee (*A. cerana*) rearing was initiated intensively by the Department of Agriculture (DOA), Parit Botak, Johor, which was led by a former agriculture officer, Mr. Zakbah Mian, in 1988. However, the practice had faced the biggest challenge in Malaysian Beekeeping history. From the end of the 1990s to the early 2000s, the bee industry collapsed due to the mite, *Varroa destructor* infestation and sacbrood viral infection. It led to the colony collapse disorder (CCD) caused by the beekeeping activity and honey industry collapse for almost ten years.

With the Malaysian Agricultural Research and Development Institute (MARDI) initiative and DOA, Malaysia, the stingless bee was introduced in 2008 as a natural pollinator agent to overcome problems related to the honeybee species. Many beekeepers prefer to rear stingless bees instead of honeybees as the stingless bee rearing technique is easier and approachable. Also, honey produced by the bees has a unique taste and aroma, plus various medical claims, which attracts the market's high demand.

In 2020, both beekeeping related to the honeybee (also known as apiculture) and the stingless bee (meliponiculture) is still available in Malaysia. Even though more attention has been given to stingless beekeeping, various activities have been strategized to recover honeybee rearing activity. It ensures both beekeeping practices can be commercialized since both bees produce distinguished and unique products and roles.

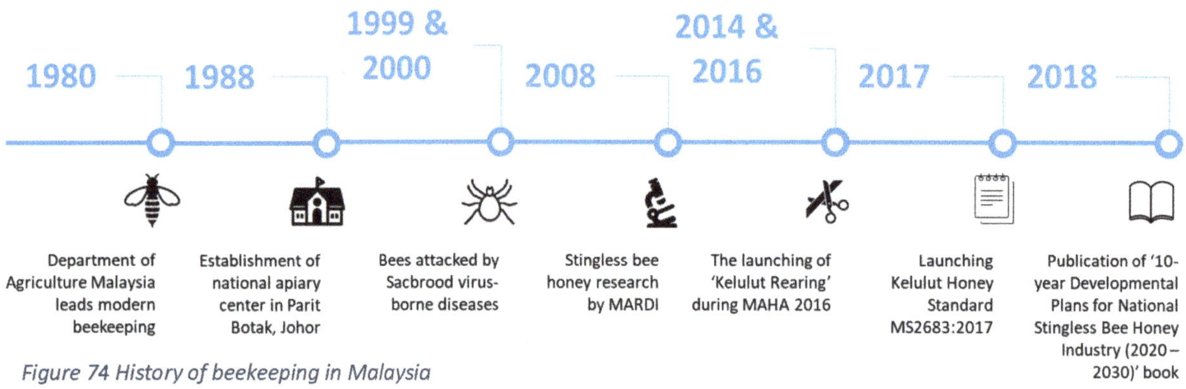

Figure 74 History of beekeeping in Malaysia

14.3 Honeybees and Stingless Bees

Honeybee and stingless bee share the same family taxonomy, which they are classified under family: Apidae, class: Insecta and Kingdom: Animalia. Both of the bees are used in the Malaysian beekeeping sector for pollination services and honey production.

However, they have prominent differences; for example, the honeybee is well-known for its stinging capability. The stingless bee lacks a stinger, which makes stingless bee preferable to be applied in local beekeeping. Moreover, the bees have a different hive structure[71], where the honeybee builds a comb-based hive, while the stingless bee forms a pot-structure hive. At least five honeybee species are available in Malaysia, such as *A. dorsata, A. cerana, A. florea* and *A. mellifera*. Meanwhile, more than 40 species of stingless bees have been discovered in Malaysia, with four species are commercially cultivated: *Heterotrigona itama, Geniotrigona thoracica, Lepidotrigona terminata*, and *Tetragonula laeviceps*. *Heterotrigona itama* is the most common stingless bee due to its higher honey production than other stingless bees.

14.4 Bee Products

Honey produced by honeybees and stingless bees has significant differences, especially from sweet and high viscosity of honeybee honey to sweet-sour and low viscosity of stingless bee honey. Besides, chemical compositions in the stingless bee honey are higher than honey harvested from the honeybee. For example, the sugar of trehalulose is only found in stingless bee honey and not in honeybee honey. More studies need to be conducted to reveal the assumption.

Editor's notes

[71] honeybees build combs and stingless bees build pots. [a colony is the bees themselves; the nest is what has a 'structure']

Figure 75 Bee products

Besides honey, both honeybees and stingless bees produce wax and propolis. They also collect pollen from floral sources and store[72] it in the brood combs (for honeybees, producing bee bread) or in storage pots (for stingless bees, producing pot-pollen). After mastication by the bees, the pollen is transferred into the brood cells with other ingredients, including enzymes and microbes, from the bee's saliva.

In terms of propolis, stingless bees produce more propolis than honeybees. Most of the stingless bee's hives are constructed from propolis, while honeycomb is mostly from wax. Unlike honeybees which build their hives primarily out of wax, most stingless bees incorporate plant resins and gums to become propolis for their nest construction.

Besides honey, bee pollen, bee bread and propolis, honeybees also produce royal jelly and beeswax. Royal jelly is a special diet prepared by honeybees to feed the queen bees and their young. For stingless bees, the queen's provision is extra masticated pot-pollen in lieu of royal jelly. Royal jelly is a mixture of natural products secreted from the hypopharyngeal and mandibular glands of nurse bees). It contains various nutrients such as proteins, amino acids, lipids, carbohydrates, and vitamins. 10-hydroxy-2-decanoic acid (10-HDA) is the most known bioactive component of royal jelly plays important roles

[72] Editor's note: Honeybees store pollen. In the brood cell, it becomes 'bee bread'. It acquires microbes and ripens in both places, but they are NOT the same.

in various biological activities, including inflammation and oxidative stress. Royal jelly has a high commercial value than any bee product. However, the amount of royal jelly produced in a hive is minimal and meagre at best by stingless bees.

Similar to bee pollen, honeybees can produce more beeswax than stingless bees. Beeswax from stingless bees is mixed with propolis, which is required a special extraction method to extract it from their honey pot. Meanwhile, beeswax from honey is more abundantly and can be extracted easily.

Lastly, a product that is only available in honeybees is bee venom. Honeybee stings and produces venom from the process. The bee venom has been used for many years in apitherapy, known as apipuncture. The apipuncture practice is similar to acupuncture, using the sting and the venom from bees, producing more medicinal health benefits than standard acupuncture techniques.

14.5 Honey Industry in Malaysia

The honey industry in Malaysia has been growing rapidly in the past decade. However, Malaysia is still far behind other countries, including ASEAN countries like Vietnam, Myanmar and Thailand. The honey industry in Malaysia is not well explored and expended less than 5% (98 mt/year) produced in 2000, but it started to increase where 8548 mt of honey was produced in 2010. In 2015, the highest production weight for honey in Malaysia was 13,767 mt, and the highest consumption weight was 15,159 mt. Although the production numbers are increased drastically throughout the decade, honey's demand always exceeds the total honey produced. This situation reflects the insufficient local supply to meet the domestic demand for natural honey, thus increasing imported honey from other countries. However, the purity of imported honey is questionable.

Nowadays, many initiatives are being conducted by authorities with various parties to promote Malaysia's honey industry. One of them is to highlight the use of local bee species (stingless bees, *A. dorsata* and *A. cerana*) compared to imported species (*A. mellifera*) for commercialization. Another example is to promote honey as a food supplement. This is because Malaysian consume honey for medicinal purposes, which mean the honey intake is minimal throughout the year or only during the sickness period of less than a week. For this reason, the honey must be pure and not adulterated.

Based on the Stingless Bee Industry National Developmental Plan 2020 – 2030, approximately RM 22 million (RM 19,269,286.00 for 134,244.49 mt of stingless bee honey RM 2,698,846.00 for 27,730 mt of honeybee honey) was gained from the honey industry in Malaysia. Sarawak contributed about 51.5% to this honey production. The value is based on 605 stingless beekeepers and 651 beekeepers registered under DOA in 2017. However, it is suggested that the value is far behind the actual figure as not all beekeepers registered under DOA and not all report their revenues since the report is a volunteer-based practice. For example, no revenues were reported by beekeepers in Sabah for honey harvested from honeybee in 2017, even though Sabah has 618 beekeepers.

Moreover, it is reported a small volume of honey (155 kg of stingless bee honey) were sold in Sabah. In contrast, Sabah is one of the bigger producers of stingless bee honey in Malaysia. Based on our random survey, the number of stingless beekeepers in Malaysia can be more than 1000. Thus, more efforts should be organized to get the actual data to make the most precise target and plan for better beekeeping development in Malaysia.

14.6 Sustainable Beekeeping and Challenges

The product from the Malaysian honey industry through apiculture and meliponiculture has the potential to become a new commodity. It can be used to reduce rural and urban poverty. Various plans have been designed in the 10-year Developmental Plans for National Stingless Bee Honey Industry (2020 – 2030). Overall, the plans are focused on the following issues: Improve governance structure related to a local beekeeping

1. Increase research and innovation
2. Increase number of experts
3. Expand business and new supply chain
4. Focus on the economic balance
5. Increase the quality and warrant authenticity status of honey
6. Use bees as pollinator agents in the agriculture setting
7. Introduce and acknowledge beekeeping as a new industry to produce a new commodity

The Malaysian honey industry is really in need of an export market. To remain sustainable, the honey, especially from the stingless bees, must penetrate the international market. The current local market is small and cannot sustain the honey industry in the long run. This is because the cost of production keeps on increasing every year. Furthermore, poor management practices lead to colony absconding affect production cost, limiting the industry expansion.

Based on honey production, the maximum potential output is still not achievable. More colonies are needed to harvest nectar from vast tropical forests, Gelam forests, and fruits and vegetables. The country is still in need of *Apis* species to do the job. Stingless bees are only able to collect a minimal amount of nectar. Malaysia should focus on both honeybees and stingless bees to maximize nectar collections. Huge land areas planted with palm oil and rubber should also be exploited for colony expansion to help smallholders during low palm oil and rubber prices. Honey production enhances smallholders' income and improves their daily nutritional intake through their honey consumption. Government agencies that manage smallholders should encourage them to plant a plant preferable by the bees like *Averhhoa carambola*, *Antigonon leptopus*, *Ocimum basilicum* and many more along plantation borders. This bee-friendly effort will help to improve biodiversity in the mono-crop areas and, at the same time, refute the anti-palm oil campaign on deforestation.

14.7 Future Outlook

In Malaysia, stingless bee honey is promoted as a superfood. It has been recognized based on its high medicinal value and effects compared to other natural products. Also, bee products from *Apis* species offer more commercial value to be expended. Malaysia is known for its epicentre of biodiversity for plants, which will give extra benefits in producing a high premium quality of honey and other bee products. However, more efforts need to be planned from all stakeholders in the beekeeping and honey industry to place local honey from both species of stinging bees (*Apis* spp.) and stingless bees (Meliponini spp.) at the international level.

14.8 Reference

1. Hassan, Z. (2003). Aktiviti Keusahawanan Masyarakat Melayu pada Zaman Kegemilangan Kesultanan Melayu Melaka. In Seminar Za'ba Mengenai Alam Melayu (Nilai-nilai Peradaban Melayu). 1-21.
2. Heard, T. A. (1999). The Role of Stingless Bees in Crop Pollination. Annual Review of Entomology, 44(131): 183-206.
3. Idris, S. H. (2013). Tingkatkan Penghasilan Madu Lebah Tempatan. Agrobiz, 6.
4. Ismail, M. M. (2008a). Gelam Conservation through Honey Hunting and Ecotourism in Terengganu. The Malaysian Journal of Agricultural Economics, 21.
5. Ismail, M. M. (2008b). Production and Trade of Natural Honey in Malaysia. In Proceedings of the 29th MSAP.
6. Ismail, M. M. (2012). Honey Marketing in Malaysia, Agricultural Marketing Issues of Selected. Commodities
7. Ismail, M. M. (2014). Competitiveness of Beekeeping Industry in Malaysia. Inaugural Lecture Series, Universiti Putra Malaysia.
8. Mardan, M. (2008). Connecting the Bee Dots. Inaugural Lecture Series, Universiti Putra Malaysia.
9. Mian, Z. (2009). Status Perlebahan di Malaysia. In Seminar Teknikal Jabatan Pertanian Malaysia.
10. Ismail, M.M. and Ismail, W. I. W. (2018). Development of stingless beekeeping projects in Malaysia. Web of References, 52(2018). DOI: https://doi.org/10.1051/ e3sconf/20185200028
11. Marlia Musa, Mohd Mansor Ismail, Wan Iryani Wan Ismail and Elpawati. (2019). Effectiveness of Extension Agent Services in Influencing the Adoption of Modern Hive in Sustainable Stingless Beekeeping. Journal of Sustainability Science and Management. Volume 14 Number 4, August 2019: 14-24
12. National Stingless Bee Industry Development Plan 2020-2030. (2019). In Malays Version.

Chapter 15

15.0 Learning from Failures in Meliponiculture and Improper Strategies of Stingless Beekeeping in Malaysia

By *Abu Hassan Jalil1 & Jeffry Iskandar Foo1

Corresponding email: *abuhsn@gmail.com

[1]Akademi Kelulut Malaysia Sdn Bhd, 2625, Persiaran Rahmat, Teras Jernang, 4350 Bangi, Selangor, Malaysia.

15.1 Synopsis

In a newly emerging industry, the maintenance and farming of stingless bees have the potential to be developed as not only a source of income for smallholders but also as a commercially expandable business. This paper explores the factors that may lead to failures and collapse in stingless beekeeping and farming as experienced in some parts of Malaysia. Many breeders, especially those who are just trying it out and newly involved in this industry, often face some challenges. Meliponiculture entrepreneurs' competency is low in terms of skills, knowledge, and entrepreneurship in the rearing of stingless bees. Lack of consolidated studies or research on local stingless bee breeding and the bee's diversity could be a reference or guide. There is poor awareness of the threats and deterioration of the surrounding ecology caused by reckless hive hunters and careless loggers. Methods of overcoming these setbacks are suggested herein.

This paper includes a list of valid species and their distribution in Malaysia, indicating the groups that are dependent on Dipterocarp resin. The vulnerability of these peculiar groups is related to their building resource trees. Besides the uncontrolled depletion of these valuable trees by hive hunters, the current unsustainable use of the forest resources inevitably leads to habitat alteration and fragmentation, which can lead to species extinction. Considering the ongoing threats to the region's biodiversity, there are concerns that many of the plants and wildlife species may completely disappear even before they are known.

Keywords: meliponine, dipterocarp, Indo-Malaya, crossing borders, hive transit, cavity volume, resin forage.

15.2 Introduction

Despite the excitement and keen interest in meliponiculture, there are many important factors to consider in promoting this industry's success. Failure in this meliponiculture endeavour may occur due to various factors. Breeders who are interested in starting meliponiculture as a small hobby right up to those who want to do stingless bee rearing at commercial farm levels need to be aware of the importance of these factors. Each of the factors discussed below is among the key issues that need consideration to ensure the industry's continuity and sustainability over the long term.

15.3 The Meliponiculture Industry Is a New Industry

As a newly emerging industry, the maintenance and farming of stingless bees have the potential to be developed as not only a source of income for smallholders but also as a commercially expandable business.

However, due to the current age of the industry, these are the main problems common to breeders:

15.3.1 Lack of information.

The reality of establishing this new industry in the country constitutes a stumbling block related to the need for information and technology transfer. The main problem faced by many breeders and 'enthusiasts' or 'hobbyists' is the lack of proper and authentic information.

• Much of the information[73] on social media sites or in mass media is often inaccurate and sometimes conflicting with each other.

• Most of the information put up has no scientific validity and no evidence-based but is merely based on experience, testimonial or trial and error of breeders. Generally, these sharing breeders do not have much experience in the industry. Because of this, lots of shared information is not necessarily correct and valid. At present, there is no body or establishment responsible for monitoring and regulating this information.

Editor's notes

[73] Much of the info on social media sites is, upon closer inspection, a testimony with no further evidence. This, unfortunately, is part of the general problem for establishing 'evidence-based' guidelines and practices. Without the evidence, or some facts that offer a test of the given conclusions, one is forced to 'fly blind'.

At present most breeders or those interested in rearing get the information from several sources as follows:

• Main source = internet's worldwide web.
• Impromptu workshops
• Club and informal associations or interest groups in social media
• Friends and neighbours
• Published books
• Government bodies such as MARDI, RISDA
• Lectures in Universities
• Private establishments such as academies and experienced beekeepers offer courses related to breeding farms with or without accreditation.
• Formal Cooperatives and Associations of state breeders
• Log (wild colonies) traders

One way to deal with this problem is to introduce a special centre responsible for collecting and filtering all information on agglutination from reliable sources like research institutions and higher learning institutions and even from successful breeders. Solution: Integrated Meliponiculture Information System. (e.g. Reliable One-Stop Information Centre for Meliponiculture.)[74]

The following are common problems that lead to inconsistencies and colony establishment failures.

15.3.2 Procurement – The type or nature of the colonies taken and removed from the habitat of origin:

The first important factor in ensuring successful meliponiculture is 'procurement' or log (wild colony) acquisition. Felling logs correctly to provide the appropriate species for breeding (whether farmed in urban or suburban areas or only suitable for forest areas) is the first factor to consider before a log/colony is retrieved for breeding purposes.

Among the problems that occur are uncontrolled logging from forest reserve areas or secondary jungles in general, especially wild bee species that mostly require resin sources obtained from dipterocarp forests. New breeders and enthusiasts who lack knowledge in terms of species identification hastily buy logs (feral colonies), and such logs go with the intention of keeping them in urban or subdistricts.

Editor's notes

[74] Any examples in other natural resource industries? It sounds like ordinary apiculture which, for some reason, initiated IBRA in the UK and now continues with an Eva Crane Foundation, but nobody, as far as I am aware, is trying to prune down or defeat the ocean of inadequate, partial, biased, untrue or poorly conceived science or information. Yes, I include 'science' because a lot of the published scientific papers include things that range from stupid to brilliant.

The buying or collection centres for those feral colonies do not have foraging sources provided (see images Figure 76 & 82), for such situations always lead to the colonies' death. And if not stopped, it may cause the extinction of these 'treasures' from the forests.

15.3.3 Felling / collecting colonies at the wrong time

Stingless bees have the forging habit of venturing out in the mornings to find nectar sources, pollen and resin for nest construction and usually return[75] to the nest/colony before sunset. In the morning or midday, most of the population will be outside the nest for the activities mentioned above. If a wild colony withdrawal is made when most of the bees are not in the nest, then the nest is vulnerable due to the reduced numbers. When considering the stress factor experienced by a colony during the transfer from the original habitat to the breeding grounds, many transplanted colonies cannot survive long and collapse within a short time after being transferred[76].

The log collection centre where wild colonies are sold to hot[77] (excited) and eager customers clearly show a lack of food sources for the bees in transit.

Figure 76 The log collection centre where wild colonies are sold to hot and eager customers, clearly shows a lack of food sources for the bees in transit.

Editor's notes

[75] Actually, they come and go multiple times during the day. [I have a large unpublished study on one exemplary species in Latin America. The colony has 7000 bees in the air at any time, an individual makes a mean of 7 foraging trips per day- in the species I studied, they were not 'time-limited' as to when they forage. They work for about 12 h a day].

[76] One of the main reasons is likely that many of the adult bees remained in the field and were not transported with the nest.

[77] this has a double meaning. Physically hot or wired and ready for 'action' of various kinds...

15.3.4 Getting colonies from across borders

Figure 77 Logs at newly arrived at transit stations with absolutely no resource provided

As a result of human greed and due to high demand, a study by Mr. Fahimie of MARDI, the species of *Heterotrigona* (*Heterotrigona*) *itama* at this time has experienced extinction in the state of Kelantan. This unfortunate predicament will happen to other states throughout Malaysia if a sustainable breeding method and proper knowledge are not disseminated,

Currently, some individuals take the opportunity to carry/colonize colonies from neighbouring borders like South Thailand and nearby Islands of Indonesia at the Straits of Malacca. The taking of colonies from the outside leads to some problems. The first problem is in terms of the risk of disease recognition (Biosecurity issues).

At present, there are no diseases associated[78] with wild species in this country. This factor is one of the advantages of endemic breeding species in Malaysia. Therefore, the introduction of other foreign

Editor's notes

[78] need a stronger statement- hive beetles *Aethina* in Africa were pests and were recognized only after someone who was working on them (as a biologist) noticed that colonies that were far from normal were being attacked. ANY OTHER NATURAL enemy or pathogen might also appear (in Malaysia, Australia, S. America, wherever) only after a number of colonies are around that really are abnormal and are susceptible to them. One of my points here is that those natural enemies were there all along. Nobody saw them in healthy colonies.

from Malaysia increases the risk of disease recognition that has not existed for local species in the country. Additionally, with uncontrolled colonization, lack of knowledge in the breeding of stingless bees and now faces the risk of illness or weak state to defend against impending threats, it is worried[79] that the country faces a problem with extinction.

15.3.5 Smuggling logs from neighbouring countries.

There are also cases where colonies smuggled out of Thailand, or other foreign countries were burned after confiscation when seized by the enforcement parties (usually Government Border security). Although bound by Malaysia's laws, unfortunately, the victim is the eager naive keeper and the bees themselves. (Figure 78) This will also contribute to the extinction of the world's chronic state of bees. Therefore, a better way of handling should be sought to monitor the smuggling problem, and a new modus operandi needs to be introduced by enforcement agencies.

15.3.6 Incorrect cutting procedure

The trees having feral colonies are felled in unscrupulous ways (severed brood cells etc.). The wrongly chopped trunks do not allow the colonies taken to be viable due to damage to the colony's shields (batumen shell lining) and the brood cells of the colony[80].

15.3.7 Acquired colony at daytime - Transfer of colony during the day/heat

Among the problems that result during wild nests retrieval during the day were risks where the nests are exposed to heat elements that could add to stress and subsequent weaknesses of the colonies. Similar to us humans, bees also take time to adapt to new environments. If the transferred colonies are too weak due to lack of numbers (population) and exposed to heat, this can lead to failure and collapse of colonies. (Figure 81)

Editor's notes

[79] I find this to be sensationalism. First, it does not follow that anything disappears just because a lot of people look for it and do damage to the environment. Second, any harm is potentially countered by natural selection. Third, you are implying that other bees from somewhere else harbour and carry diseases into Malaysia. Watch this, very carefully, because it has two problems: 1) diseases usually are species-specific, incapable of jumping from species to species and 2) your thesis seems to be that everything would be just fine in Malaysia if no colonies were imported.

[80] These are completely normal if the local bee hunters have no idea what the buyers want or what bee biology is about. They usually would assume that what was in the nest- honey, pollen brood, was what was being purchased. Who cares about the adult bees and foragers? Even the queen has no meaning to such people. They are ignorant.

15.3.8 Logs in Transit

The log hives in transit are often wrapped with plastic insulation tape to prevent rain exposure, but this only aggravates matters with heat build-up[81]. On top of this, the trailer covered with tarpaulin accumulates heat during a sunny day and chances are 50% of the colonies have suffered irreparable damage. (Figure 78)

Alternatively, the preferred method is to wrap in a cotton jersey and transport them in reasonable quantities in air-conditioned vans. (Figure 79)

Figure 78 Recklessly stacked of newly smuggled log hives, poorly prepared for quick colony disposal to buyers.

Editor's notes

[81] asphyxiation occurs- No oxygen, carbon dioxide build-up, just like humans

Figure 81 Transporting log logs in midday under the hot sun.

Figure 80 Newly cut log hives poorly wrapped and stacked onto a trailer

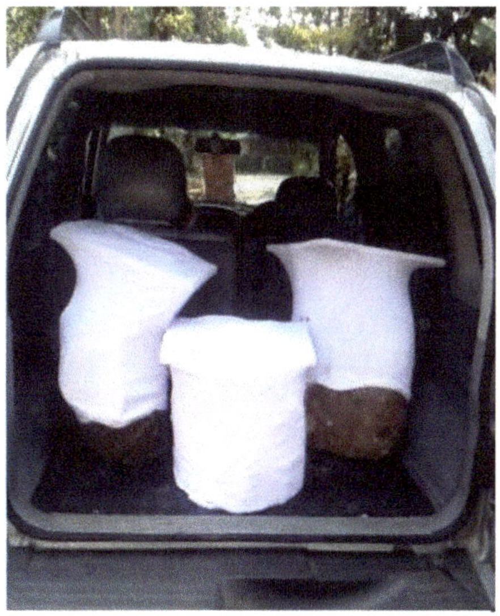

Figure 79 Preferred wrapping material

15.3.9 Size of log nest

Bees make nests as long as there is a cavity in a tree, regardless of tree trunk size[82]. Mature rainforest trees naturally have cavities when the internal xylem outgrows its volume. These cavities are accessible through branch stubs that have rotted away. When this happens, chances are the trees are fully grown, and their girth can be more than one meter and most times approaching 2 meters. (**Error! Reference source not found.**)

Figure 82 Reckless cutting and shaving of log hives.

In their excitement, hive hunters or log fellers only think of the method of extraction after cutting down the tree. Ending with a massive log, they start slicing and shaving the sapwood so that the whole mass can be more manageable for extraction to awaiting transportation outside the forest[83]. More often than not, this shaving and slicing creep into the nest proper, thereby damaging the whole colony.

Editor's notes

[82] They will normally be found where the cavity volume is suitable, and that is highly correlated with the size of the tree. There is an acceptable range of cavity width for most species, and it can be surprisingly narrow- 5-8 cm. But less than this makes transit of air, the queen, circulation of carbon dioxide and other things impossible, depending on the actual architecture of the brood area and involucrum. Some things must take place in a cavity that is fairly wide, and that does not exist in slender trees or branches. Bees with brood arranged in loose clusters can much more easily use narrow cavities- as you have seen.

Editor's notes

[83] If they cut out brood, this is bad. If they remove honey or pollen, this is better than having a large log that cannot be transported. They mainly have to be careful to close the exposed portions of the nest, and to leave the nest in situ long enough for field bees to return, and to enter the nest in its altered place and condition. Then, they must PLUG UP the nest entrance, so that the adult bees STAY INSIDE. This also depends on preserving the nest entrance intact, or in putting the old nest entrance (which is cut off) over the new entrance to the transported, cut log hive. Not anyone can do this, to be sure, but most people figure it out pretty quickly. It may require waiting until dusk, when most of the field bees are searching intently for their nest, which they usually find, if it is not more than several m from its former 3D position.

Figure 84 Farming in close quarters and narrow residential areas—livestock in low lying areas prone to flooding by the unprepared hobbyists.

Figure 83 The farm area is clean but too dense and hot.

15.3.10 Hive & Stock location

As mentioned earlier, many wild colonies consist of species that can only be reared in the forest (*Tetrigona* spp.) and should not be farmed in urban or suburban areas.

Figure 85 Damage done by a Malaysian Sun Bear

Breeding problems in forest areas

Wild colonies were attacked by a bear.

Forest fringe areas[84] is usually a very suitable area to choose from as a breeding ground primarily because colonies are typically shielded from hot weather extremes and also because there are so many essential sources for forage available easily. However, there are still risks in this area. Among the dangers of breeding the fringe of the forests first is the threat of dangerous wild predators. In Malaysia, the most important predator is the sun bear that tries to propagate and destroy colonies with sharp, tough claws to take honey which is the food for this bear.

The forest area is also often the least inhabited. If no guardian in this area always can lead to the risk of theft.

Bear and other predators (like wild boars and aggressive alpha monkeys- short-tailed macaques) attacks can usually be managed by human control. Hence, it is essential for breeders who run livestock in such areas to ensure that there are measures taken for caring for this area to ensure no intrusion from wild animal predators and theft.

15.4 Problems faced when farming in urban or suburban areas

Today there are many enthusiasts or hobbyists who breed bees as pets in a residential area. Among the issues faced by this area is the food source for the colony. In the excitement, they may transfer up to 10 colonies at a time (concurrently) in the city's residential area. The transfer of such colonies simultaneously to unsuitable locations (lack of resources and heat issues) causes all the colonies destroyed within a short period.

The second factor is the climate or temperature. As has been already stated above, the overheated weather is not suitable for breeding.

Editor's notes:

[84] You mean a Meliponary near intact forest. A forest fringe can also be within forest. They are two different things, but you are right, the predators are not living in the corn fields or palm plantations.

15.4.1 The use of herbicides and mosquito fogging.

A lack of resources is the colony's exposure to herbicides/pesticides in urban or rural agriculture areas. Among the familiar sources of forage for the bees are weed flowers, and the herbicides like Round-up eradicate these.

The use of herbicides causes the bees to take the toxic contaminated sources back to the colony. This cause adverse effects[85] to the next generation of bees and, in the long run, will lead to the destruction of the colony.

15.4.2 Bee deaths due to herbicides/insecticides.

Fogging (malathion) for Aedes mosquitoes that usually occur primarily in residential or suburban areas exposes the risk of colony death.

Figure 86 Beekeepers need to avoid placement of stock in the area of frequent fogging of mosquitoes.

15.4.2.1 Methods of overcoming mosquito fogging

Proper distribution of information to the public to create awareness through mass media, print media, internet etc.

- Conduct meaningful communication between neighbours and breeders.

Editor's notes:

[85] If it is a toxin, it kills them, and if it is a mutagen, it may hinder development or make them inviable or kill them. If it is a neonicotinoid, like glyphosate and imidacloprid, it might not kill them, but make them incapable of memory and orientation needed to forage, mate or return to the nest.

- Allow communication[86] between breeders and the District Health Office.
- Send notifications to the breeders by the end of the week before fogging is conducted so that the breeder can take appropriate preventive measures.

Residential areas where fogging is a routine;

- Beekeepers can cover their hives with damp burlap or discarded blankets.
- The beekeeper needs to spray water on the nearby vegetation (especially those allocated as a foraging source for the bees). This will reduce the fogging chemicals' residue and prevent them from adhering to the flowers and eventually being absorbed by the pollen.
- Water before the intended fogging exercise and then after the fogging before the next morning when the bees will be going out foraging.

Figure 87 This topping box is too big.

15.4.3 Improper strategies in breeding farms

Many breeders believe that large topping boxes can lead to a high volume of honey production. Unfortunately, this idea is false as the beehives require the need for housekeeping or nest cleansing. Space occupied that is too large requires the use of much energy and more effort to be maintained. Likewise, beehives limit their production (capacity and product volume) to colony size and population. Therefore, preparing standardized volume hive boxes without considering colony size is unfounded.

Many fierce colonies with large-sized hive boxes will have humidity problems, poor temperature regulation and subsequently pest intrusion. A couple of oversized topping boxes (a.k.a. "honey supers" as in Apiculture)[87] with weak colonies (weak due to transferring to a new location). The colony tries to adapt to the new environment, and this will add stress to a colony.

Editor's notes:

[86] Include the pesticide industry or integrated pest management people? Both may be around, or at least the former. No, they usually will not help to preserve non-target wildlife...

Editor's notes:

[87] big supers lead to problems that are unnecessary. Why not just wait to add the honey super until the colony is stable?

15.4.4 The proposed volume of hive boxes

As a guide, these are cavity sizes that occur in nature by species.

Body length, cavity Volume and the average population of some *Stingless Bee*

1. Tetragonula. minangkabau: body length 3.2 – 3.5 mm. volume 0.5 litre, average population 2600 bees.

2. *T. drescheri:* body length 4.8 – 4.9 mm. volume 1.4 litre mature nest volume was 1350 cm3 to 7490 cm3

3. *T. laeviceps:* body length 4.0 – 4.5 mm. volume 2.0 litre

4. *Heterotrigiona itama:* wing length 6.2 mm. volume 4.5 litre, population avg. 5000

5. *Sundatrigona. moorei:* wing length 4.6 mm. population avg. 1800

6. *Wallacetrigona. incisa: body length* 8.6 ± 0.4 mm

Source: Siti Salmah Idrus, National Bee Seminar 2017, IPB, Bogor, Indonesia

15.4.5 Diverse selection

Having a varied and diverse selection of meliponines has its advantages and disadvantages.

On the upside, beekeepers have an excellent choice of bees for farming to suit the natural habitat available (albeit with complete knowledge of which species having a preference for which type of vegetation).

On the downside, one cannot provide for a standardized box hive or 'topping' box because the different species have a preference for various sized cavities.

Occasionally we encounter beekeepers who insistently provide a standard size (economics of mass box hive production). These forced breeding practices may be detrimental since a forced population is a stressed population. Production may never seem to be optimized.

15.4.6 Exposure of paints in beekeeping

Figure 88 "Creative' keepers painting boxes to preserve the hive boxes but using the wrong type of paint

At present, there are 'creative' breeders in terms of beautifying the boxed beehives by painting the 'topping' boxes. Those who know about the bees usually see that they like to take resin ingredients or terpene-based sources. Most paints contain lead-based components and resinous compounds. Stingless bees are attracted to paint that has not entirely dried. They have an affinity to gnaw[88] at the paint's resin, especially when the natural vegetative resin is in shortage. These resins are commonly used to build nests, including honey pots. Problems arise because paints also typically contain heavy metals such as lead that are harmful to humans. The heavy metal pollution will damage the quality of honey produced.

Recommended paints, if hive coating is necessary, are rubber-based Industrial paints or water-based whitewash. Most glossy paints[89] used in Malaysia have heavy metal components that may be detrimental to the bees and may contaminate the honey produced.

Editor's notes:

[88] They can't drink the paint. They are chewing it up to transport it, on the corbiculae, to use as building material

[89] Glossy acrylic paints are the norm for interior and exterior paints in the USA. They also come in semi-gloss and matte. It has been this way for over 20 years. Lead paints have been outlawed.

Figure 89 recommended acrylic paint finish – before & after

Figure 90 Hive boxes finished in water-based whitewash from late 2000 still going strong to date.

15.5 Species Distribution

Considering the diverse types, it is no wonder that the peculiarities and variations in foraging habits make for a complicated operating procedure for Meliponiculture in South East Asia.

Types of *Stingless Bee*

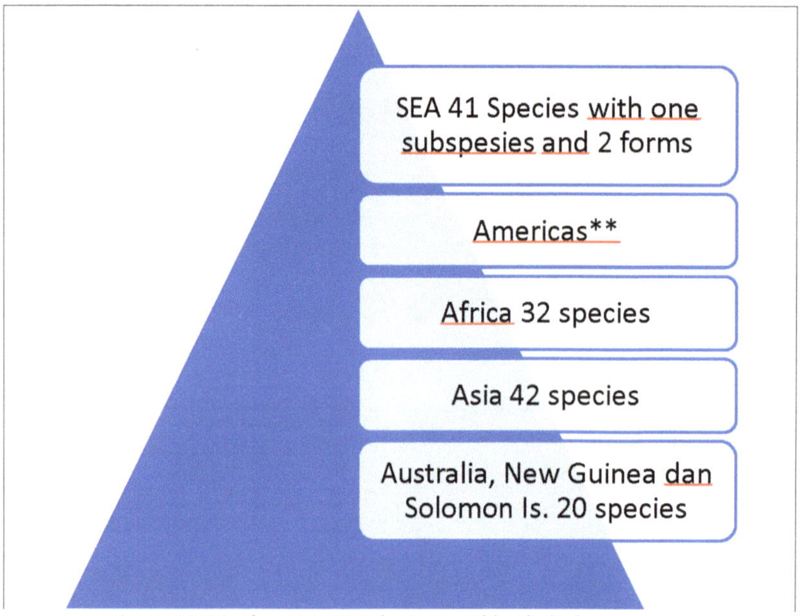

Figure 91 Stingless Bee Distribution worldwide (Schwarz 1939)

In Southeast Asia (Schwarz, 1939)

Borneo 31 species

Sumatra 18 species

Java 9 species

Malaya Peninsula 29 species

Thailand 20 species

Burma 9 species

Indochina 8 species

Philippines 6 species

**[90]

Overlapping East Malaysia (North and NW Part of Borneo) and West Malaysia (Malay Peninsula), we have a total of 48 valid named species in Malaysia (Deborah Smith, 2012)

The distribution of stingless bees in the Indo-Malaya/Australasian region stretches from India to the Solomon Islands and from China (Yunnan, Hainan, Taiwan: Wu 2000) to Australia (New South Wales). Schwarz (1937) suggested that the greater abundance of stingless bees in Thailand and Malaysia, including all of Borneo, is due to the abundance of resin-secreting trees (Dipterocarpaceae) and humid tropical climate. However, most likely, an array of factors is responsible for this concentration of species.

Editor's notes:

[90] Easily 400 species in S. America... and 100 in Central America and Mexico. This is documented, most recently by me and by Grúter (2020 book). This is the number of valid species, not the number of unpublished, new species. We know, roughly, how many in museums have no scientific names. In Brazil alone there are about 340, 412 in tropical America, and this was recorded and published in 2013 and later- Camargo, Pedro, and others.

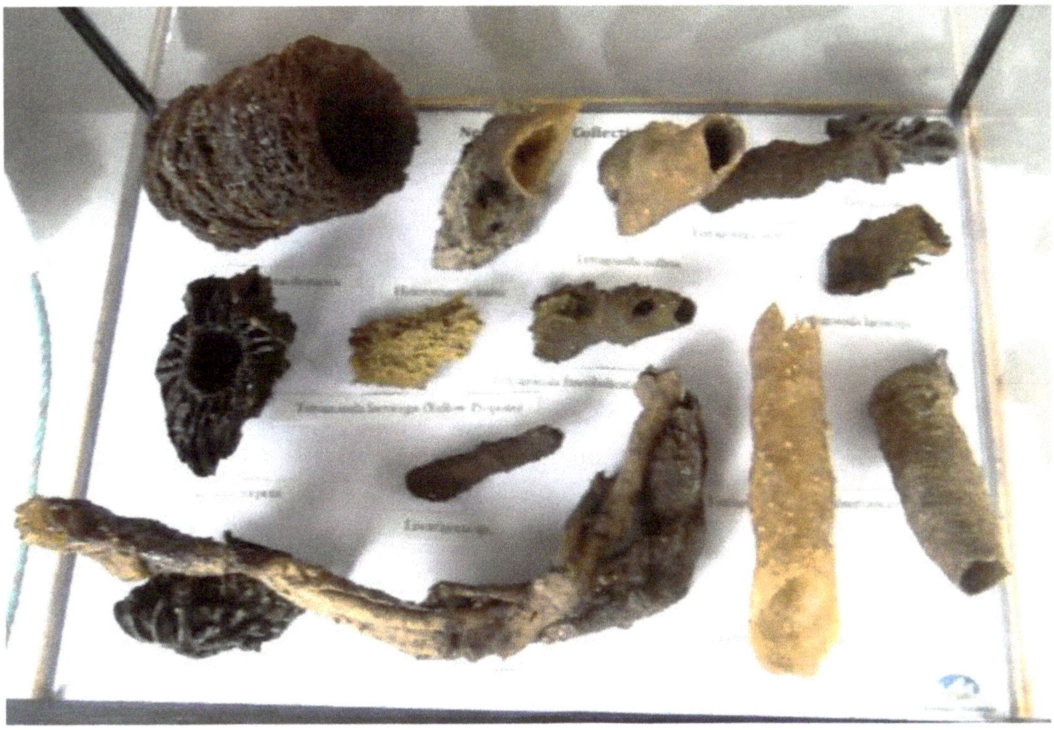

Figure 92 Selection of diverse nest entrance tubes - Hard resinous tubes is Dipterocarp resin sourced.

15.6 Dipterocarpaceae in Malaysia - Endemic Resin producing Trees

Figure 93 Dipterocarp winged seeds

Description	Permai Rainforest Resort, Santubong Peninsula North of Kuching, Sarawak, MALAYSIA
Date	3 October 2014, 05:47
Source	Dipterocarp Winged Seeds (Shorea sp.)
Author	Bernard DUPONT from FRANCE

Ref: https://en.wikipedia.org/wiki/Dipterocarpaceae accessed 17 February 2018

Figure 94 Engkabang (Shorea sp.)

Figure 95 Sweet Shorea (Shorea roxburghii G. Don)

Description Niah NP, Sarawak, MALAYSIA

Date 22 September 2014, 04:10

Source Engkabang (Shorea sp.)

Author Bernard DUPONT from FRANCE

Camera location 3° 49′ 215.88″ N, 113° 46′ 015.42″ E

Sweet Shorea (*Shorea roxburghii* G. Don) tree planted by King Bhumiphol (Rama IX) of Thailand on Khao Tha Phet, Surat Thani. The photo was taken by User: Ahoerstemeier on January 18, 2005.

Ref: https://en.wikipedia.org/wiki/Shorea accessed 10 February 2018

15.6.1 Common resinous trees with botanical and vernacular names.

Figure 96 Tress favoured by Stingless bees

Figure 97 Common Trees producing resin

The Dipterocarp dependant stingless bees have diverse alternative and peculiar resinous trees. Each group of species have to be approached with much consideration on their foraging habits and affinity.

Figure 98 Stingless Bees Commonly bred in Malaysia

Following is a list of species found in Malaysia with an asterisk denoting Dipterocarp dependent.

Table 1 List of species in Malaysia Inc. East Malaysia Borneo (Smith, 2012) (Rasmussen, 2008))

No.	Species	Ref,	Distribution
1	*Geniotrigona lacteifasciata*	(Cameron, 1902)	Malaysia- Peninsula; Borneo
2	*Geniotrigona thoracica*	(Smith, 1857)	Malaysia- Peninsula
3	*Heterotrigona bakeri*	(Cockerell, 1919)	Malaysia- Pulau Pinang
4	*Heterotrigona erythrogastra*	(Cameron, 1902)	Malaysia- Peninsula; Borneo
5	*Heterotrigona itama*	(Cockerell, 1918)	Malaysia- Peninsula
6	*Homotrigona aliceae* *	(Cockerell, 1929)	Malaysia- Peninsula
7	*Homotrigona fimbriata* *	(Smith, 1857)	Malaysia- Peninsula; Borneo

8	*Homotrigona lutea* *	(Bingham, 1897)	Malaysia- Peninsula
9	*Lepidotrigona flavibasis*	(Cockerell 1929)	Malaysia- Peninsula
10	*Lepidotrigona nitidiventris*	(Smith 1857)	Malaysia- Peninsula; Borneo
11	*Lepidotrigona terminata*	(Smith 1878)	Malaysia- Peninsula; Borneo
12	*Lepidotrigona ventralis*	(Smith 1857)	Borneo; Malaysia;
13	*Lisotrigona cacciae*	(Nurse, 1907)	Borneo; Malaysia;
14	*Lophotrigona canifrons* *	(Smith 1857)	Malaysia- Peninsula; Borneo
15	*Pariotrigona pendleburyi*	(Schwarz 1939)	Malaysia- Peninsula; Borneo
16	*Pariotrigona klossi*	(Schwarz)	Malaysia- Peninsula; Borneo
17	*Sundatrigona moorei*	(Schwarz, 1937)	Malaysia- Peninsula; Borneo
18	*Tetragonilla atripes* *	(Smith, 1857)	Malaysia- Peninsula; Borneo
19	*Tetragonilla collina* *	(Smith, 1857)	Malaysia- Peninsula
20	*Tetragonilla fuscibasis* *	(Cockerell, 1920)	Malaysia- Peninsula; Borneo
21	*Tetrigona apicalis* *	(Smith, 1857)	Malaysia- Peninsula; Borneo
22	*Tetrigona binghami* *	(Schwarz, 1937)	Borneo, Malaysia- Peninsula
23	*Tetrigona peninsularis* *	(Cockerell, 1927)	Malaysia- Peninsula
24	*Tetrigona vidua* *	(Lepeletier de Saint Fargeau, 1836)	Malaysia- Peninsula
25	*Tetragonula fuscobalteata*	(Cameron, 1908)	Malaysia- Peninsula; Borneo
26	*Tetragonula geissleri*	(Cockerell, 1918)	Malaysia- Peninsula; Borneo
27	*Tetragonula laeviceps*	(Smith, 1857)	Malaysia- Peninsula

28	*Tetragonula melina*	(Gribodo, 1893)	Borneo; Malaysia
29	*Tetragonula minangkabau*	(Sakagami & Inoue, 1985)	Malaysia- Peninsula
30	*Tetragonula minor*	(Sakagami, 1978)	Malaysia- Peninsula
31	*Tetragonula pagdeni*	. (Schwarz, 1939)	Malaysia
32	*Tetragonula pagdeniformis*	(Sakagami, 1978)	Thailand; Malaysia-Peninsula
33	*Tetragonula penangensis*	(Cockerell, 1919)	Malaysia- Pulau Pinang
34	*Tetragonula reepeni*	(Friese, 1918)	Borneo; Malaysia-Peninsula
35	*Tetragonula sirindhornae*	(Michener & Boongird 2004)	Thailand; Malaysia-Peninsula
36	*Tetragonula testaceitarsis*	(Cameron, 1901)	Malaysia- Peninsula
37	*Tetragonula zucchii*	(Sakagami, 1978)	Malaysia- Peninsula
38	*Homotrigona anamitica* *	(Friese, 1909)	Borneo
39	*Lepidotrigona latebalteata*	(Cameron 1902)	Borneo
40	*Lepidotrigona trochanterica*	(Cockerell 1920)	Borneo
41	*Odontotrigona haematoptera* *	(Cockerell 1919)	Borneo
42	*Borneotrigona hobbyi* *	(Schwarz, 1937)	Borneo
43	*Tetragonilla rufibasalis* *	(Cockerell, 1918)	Borneo
44	*Tetragonula drescheri*	(Schwarz, 1939)	Borneo
45	*Tetragonula melanocephala*	(Gribodo, 1893)	Borneo
46	*Tetragonula sarawakensis*	(Schwarz, 1937)	Borneo

*Fifteen species in Malaysia are Dipterocarp dependent, with the rest having an affinity for a varied shape or colour of flowers. (Personal observation)

15.7 Summary and Conclusion

Activities that to be avoided	Activities encouraged
Nest Acquisition *1. Removing wild colonies from their natural habitat in the daytime.*	*1. Remove the colonies from their natural habitat early in the morning before sunrise or immediately after sunset.*
2. Improper cut or shaving of logs containing colonies to reduce overall weight for transport from the original habitat can cause damage to the nests themselves. Erroneous slits and honey spills may cause the infiltration of pests.	*2. Reduce shaving of the logs and ensure no damage to the batumen nests. Understanding the environment, they are in, and each species' peculiarity is the key to successful procurement.*
3. Transport colonies from origin to collection point and then to bee farms done in the morning/evening under the hot sun, thus weakening the colony.	*3. Transport or transfer movement should be done at night.*
4. Wrapping of the log with excessive plastic bandages, causing weakness to the colony of the bees, in exceptionally long haulage.	*4. Close only the bee colony's entrance or use non-plastic wrapping materials such as cotton jerseys.*
5. Lack of resources or shading for bee colonies at the collection centre.	*5. Provide adequate resources for bees at the collection centre because the stressed or damaged colonies need to recover. Beescape the collection centre (see farm provisioning next).*

Avoid these in Farm Provisioning	***Encourage these - Providing Beescape***
1. Location is not suitable for meliponiculture. Exposure of colonies to harmful elements such as herbicides and fogging may cause colony and honey pollution failures.	*1. Find a good location, away from dangerous activities to bees such as fogging in residential areas or the use of pesticides in some agricultural regions.*
2. Location of a farm that is exposed to natural disasters such as floods or predators like bears or other wild animals.	*2. Avoid areas that are vulnerable to flooding and provide adequate protection to prevent parasites, predators, destroyers, and theft.*
3. No protection or shade from sunny high temperatures in the fields.	*3. Preparation of cool and shaded areas. The queen's functions deteriorate when temperatures*[91] *in the brood chamber rise above 29° C.*
4. Lack of vegetation and low utilization of area for trees that bees use causes a lack of food for bees.	*4. Grow enough plants that can provide food/resources/coverage for the whole year for bees.*
5. Inserting too many colonies into one location at once – need to consider colony density against species type. (As a rule, provide space in between hives at one swarm ball apart, based on the respective species size of the swarm).	*5. Do not squeeze too many colonies into small areas. Introduce a small number of colonies into the site in a staggered manner while observing the environment and monitor the colony development. Ensure adequate resources are always provided.*

[91] Bees start to die at 35 in the nest. They commonly fly in the canopy at 40 or so.

Bad Beekeeping	***Proper Meliponiculture***
1. Dirty and unkempt environment that promotes pest reproduction such as flies, which cause the failures of bee colonies and pollution of the bee products.	*1. Always maintain a clean environment. Routine housekeeping and regular farm maintenance.*
2. Improper lumber for a toping box that cannot withstand rain, heat or possible warping in ambient temperature.	*2. Choose suitable material for constructing topping boxes (hardwoods that can withstand termites and does not rot easily).*
3. Finishing of bee box with glossy lead-based paint. This will lead to the potential of heavy metal pollutants in bee products because the bees tend to be attracted to the paint resin.	*3. Avoid painting the topping box with lead-based paint but use water-based or rubber-based paint instead. The use of wax or propolis & wax ingredients may be a better option than paint. Use Acrylic paint if necessary.*
4. The unnecessary force breeding or premature splitting of the colonies without the knowledge of bee biology and sufficient beekeeping experience.	*4. Provide containers for less destructive propagation methods such as eduction box, trap box and three-tiered system. Natural reproduction should be encouraged for healthy multiplication. Provide a good food source and optimum location for colony development and propagation.*
5. Bee product pollution due to improper harvesting methods, e.g. individuals are sucking directly from honey pots using a straw. This may degrade the honey quality and increase the risk of infectious diseases transmitted between consumers.	*5. Avoid unhygienic practices such as sucking honey from the pots with a straw (backflow occurs and contaminate the remnant honey - see image below). Use gloves and a facial mask when harvesting honey (to avoid dirty hands and accidental coughing or sneezing).*

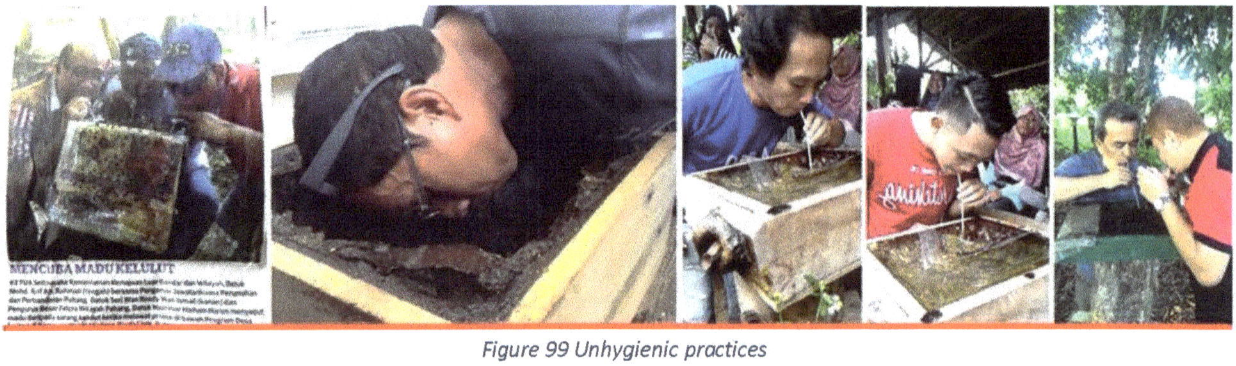

Figure 99 Unhygienic practices

Chapter 16

16.0 The Itama Journey

Nor Azfa Johari, Mohd Razif Mamat, Azira Muhamad, Nur Suhanawati Ashaari, Nurulusna Che Zain, Nurul Ain A. Talib & Azyyati Mohd Padzil

Malaysia Genome Institute, National Institutes of Biotechnology Malaysia, Ministry of Science, Technology and Innovation (MOSTI), Jalan Bangi, 43000 Kajang, Selangor

*Correspondence: azfajo@nibm.my

Listed as one of the world's megadiverse countries, Malaysia is ranked 12th globally, with estimated 15,000 species of vascular plants, 100s of mammal species, bird, amphibian, reptile, bird, marine life and more than 150,000 species of invertebrates, including insects. Stingless bees are listed as one of the most crucial forest pollinators, with 40 or more species native in Malaysia. Despite being known as home to the Indo-Malayan stingless bee, many are obscure on this species' real importance. Stingless bee is by far the largest group of eusocial bees on earth. About 800 species were recorded in entire tropical regions of the world, consisting of many ecological habitats such as dry savannah or cerrado to humid rainforest [Hrncir et al. 2016; Abd Hamid et al. 2016]. Unlike honeybees, stingless bee colonies are not uniform across species. Although the basic nest structure includes cerumen and hard resin, the entrance shape, types and resin preference are almost species-dependent.

In Malaysia, the stingless bee is considered an industry on its own. Stingless bee products have initiated a medium scale size of sector with a focus on honey production. With several advantages over Apis species, stingless bee farming or meliponiculture is preferable due to its low absconding behaviour and long-lasting colony. With strong support from the government through several ministries, industrial players have introduced many innovations on each aspect of meliponiculture, from landscape setting, hive structure, harvesting apparatus to downstream processing technology covering honey quality assessment development various by-products from Stingless bee components. The most domesticated Stingless bee species in Peninsular Malaysia are *Heterotrigona itama* and *Geniotrigona thoracica*. These two species have gained much attention from the public and industry, which is considered the most economically significant. A specific honey specification has also been developed using *H. itama* as a reference due to high accessibility and constant honey production.

On the surface, most stingless bees appear similar with the primary difference in size, but further identification needs to be carried out with a microscope platform. *H. itama* is characterized microscopically with an entirely black body, hairless mesoscutum and long malar space (Azizi et al. 2020) with aggressive behaviour upon disturbance. Despite the high similarity in physical features, the honey product characteristics can be different between colonies. A wide variety of *H. itama* honey, from colour, thickness, smell and taste, is strongly correlated with source, season, humidity and location. A one-year sampling activity for *H. itama* farm and the honey collection has been carried out in 50 locations across Peninsular Malaysia to get a baseline record on meliponiculture of *H. itama* and its honey profile for reference.

16.1 The Journey Log

Heterotrigona itama honey sampling started in June 2019 in Sungai Pelek, Sepang, Selangor, and ended on August 2020 in Permatang Badak, Kuantan, Pahang. Map 3 showed locations distribution in Peninsular Malaysia map.

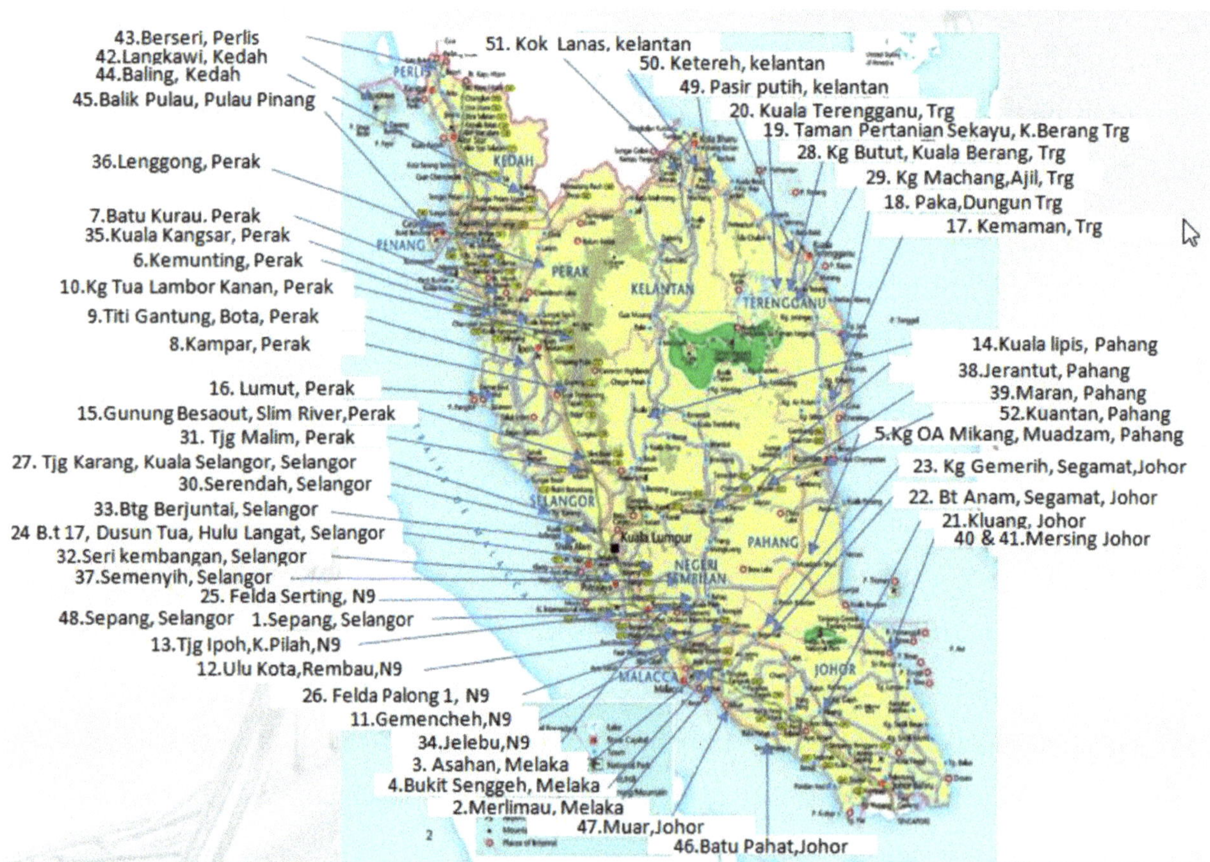

Map 3 Sampling Locations

16.1 List of Locations

Location 1 Meliponary in Sungai Pelek, Sepang, Selangor 271
Location 2 Meliponary in Kampung Sempang, Merlimau, Melaka 271
Location 3 Meliponary in Kampung Orang Asli Simpang Bekoh, Asahan, Melaka 272
Location 4 Meliponary in Felda Tun Ghafar Baba, Nyalas, Melaka 272
Location 5 Meliponary in Kampung Orang Asli Bukit Ibam, Kuala Rompin, Pahang 273
Location 6 Meliponary in Kamunting, Perak 273
Location 7 Meliponary in Batu Kurau, Perak 274
Location 8 Meliponary in Kampung Batu Berangkai, Kampar, Perak 274
Location 9 Meliponary in DOA Titi Gantung, Bota Kiri, Perak 275
Location 10 Meliponary in Kampung Lambor Kanan, Perak 275
Location 11 Meliponary in Kampung Ulu Rokan, Dangi, Gemencheh, Negeri Sembilan 276
Location 12 Meliponary in Kampung Legung Ulu, Kota, Negeri Sembilan 276
Location 13 Meliponary in Kampung Tanjung Ipoh, Negeri Sembilan 277
Location 14 Meliponary in Kampung Lubuk Kulit, Kuala Lipis, Pahang 277
Location 15 Meliponary in Felda Gunung Besout, Perak 278
Location 16 Meliponary in Lumut, Perak 278
Location 17 Meliponary in Kemaman, Terengganu 279
Location 18 Meliponary in Felda Kerteh 2, Paka, Terengganu 279
Location 19 Location 19: 280
Location 20 Meliponary in Kampung Gemuruh, Kuala Nerus, Terengganu 280
Location 21 Meliponary in Kluang, Johor 281
Location 22 Meliponary in Kampung Seri Dagang, Segamat, Johor 281
Location 23 Meliponary in Kampung Gemereh, Segamat, Johor 282
Location 24 Meliponary in Kampung Sungai Terap, Tanjung Karang, Selangor 282
Location 25 Meliponary in Batu 17, Dusun Tua, Hulu Langat, Selangor 283
Location 26 Meliponary in Felda Palong 1, Gemas, Negeri Sembilan. 283
Location 27 Meliponary in Felda Serting 4, Bandar Seri Jempol, Negeri Sembilan 284
Location 28 Meliponary in Kampung Butut, Kuala Berang, Terengganu 284
Location 29 Meliponary in Kampung Machang, Ajil, Terengganu 285
Location 30 Meliponary in D'Bullah Garden, Serendah, Selangor 285
Location 31 Meliponary in Tanjung Malim Perak, ... 286
Location 32 Apis cerana farm in Seri Kembangan, Selangor 286

Location 33 Meliponary in Selangor Fruit Valley, Selangor 287
Location 34 Meliponary in Jelebu, Negeri Sembilan 287
Location 35 Meliponary in Kuala Kangsar, Perak 288
Location 36 Meliponary in Lenggong, Perak 288
Location 37 Meliponary in Batangsi, Selangor 289
Location 38 Meliponary in Kampung Pulau Jerejak, Bukit Lada, Chenor, Pahang 289
Location 39 Meliponary in Jerantut, Pahang 290
Location 40 Meliponary in Kampung Sawah Dato, Mersing, Johor 290
Location 41 Apis mellifera farm in Mersing Bee Farm, Johor 291
Location 42 Meliponary in Langkawi, Kedah 291
Location 43 Meliponary in Perlis 292
Location 44 Meliponary in Baling, Kedah 292
Location 45 Meliponary in Bayan Lepas, Pulau Pinang 293
Location 46 Meliponary in Kampung Bintang, Batu Pahat, Johor 293
Location 47 Meliponary in Parit Sekoyan, Kesang Laut, Muar, Johor 294
Location 48 Meliponary in Bagan Lalang, Sepang, Selangor 294
Location 49 Meliponary in Kampung Semarak, Pasir Puteh, Kelantan 295
Location 50 Meliponary in Ketereh, Kelantan 295
Location 51 Meliponary in Kem Desa Pahlawan, Kok Lanas, Kelantan 296
Location 52 Meliponary in Permatang Badak, Kuantan, Pahang 296

Location 1: Meliponary in Sungai Pelek, Sepang, Selangor. Sampling began at 11:11 a.m., 28°C and in cloudy condition. The farm, owned by Nur Raziana Ramli, is located in a village area, surrounded by a palm oil plantation and a fruit farm.

Location 1 Meliponary in Sungai Pelek, Sepang, Selangor

Location 2: Meliponary in Kampung Sempang, Merlimau, Melaka. Sampling began at 9:47 a.m., 30°C and in cloudy condition. The farm, owned by Hajjah Halimah Haji Mat Johan, is located in a village near rubber and oil palm estates with multi-floral surroundings. There is also a medium-size industries area within a one km radius.

Location 2 Meliponary in Kampung Sempang, Merlimau, Melaka

Location 3: Meliponary in Kampung Orang Asli Simpang Bekoh, Asahan, Melaka. Sampling began at 12:17 a.m., 30°C and in cloudy condition. The farm, owned by Wak Aim, is located in a village near a rubber plantation with multi-floral surroundings.

Location 3 Meliponary in Kampung Orang Asli Simpang Bekoh, Asahan, Melaka

Location 4: Meliponary in Felda Tun Ghafar Baba, Nyalas, Melaka. Sampling began at 1:35 p.m., at 31°C and in cloudy condition. The farm, owned by Hairol Abdullah, is located in a village area with fruit farms and multi-floral surroundings.

Location 4 Meliponary in Felda Tun Ghafar Baba, Nyalas, Melaka

Location 5: Meliponary in Kampung Orang Asli Bukit Ibam, Kuala Rompin, Pahang. Sampling began at 2:56 p.m., at 28°C and in cloudy condition. The farm is owned by Zahar A/L Tan and is located in a village near an oil palm plantation with multi-floral surroundings.

Location 5 Meliponary in Kampung Orang Asli Bukit Ibam, Kuala Rompin, Pahang

Location 6: Meliponary in Kamunting, Perak. Sampling began at 11:48 a. m., 30°C and in cloudy condition. The farm, owned by Shahrizat Ramli, is located in the multi-floral surrounding, and stingless bee logs were placed mainly under mangosteen and durian trees. The farm is also near Air Terjun Bukit Jana, a tourism spot during the weekend. This is where the team managed to sample purple honey with a strong and intense acidic taste.

Location 6 Meliponary in Kamunting, Perak

Location 7: Meliponary in Batu Kurau, Perak. Sampling began at 2:31 p.m., at 28°C and in cloudy condition. The farm, owned by Mat Isa Haji Yeop, is located in a village area with *Areca catechu* trees and multi-floral surroundings. This is the second place where dark purple honey was sampled.

Location 7 Meliponary in Batu Kurau, Perak

Location 8: Meliponary in Kampung Batu Berangkai, Kampar, Perak Sampling began at 10:54 a.m., 28°C and in cloudy condition. The farm owned by Nordin Bahak is located in a multi-floral, dense rural area with several fruit farms, mostly durian. The farm is situated near Air Terjun Bukit Berangkai and appeared light yellow.

Location 8 Meliponary in Kampung Batu Berangkai, Kampar, Perak

Location 9: Meliponary in DOA Titi Gantung, Bota Kiri, Perak. Sampling began at 1:17 p.m., at 30°C and in cloudy condition. The farm, owned by Ikhwan Nadzir Mohamad, was located in one of DOA's fruit farm valleys, where *Averrhoa bilimbi* is the dominant species.

Location 9 Meliponary in DOA Titi Gantung, Bota Kiri, Perak

Location 10: Meliponary in Kampung Lambor Kanan, Perak. Sampling began at 4:29 p.m., at 26°C and in raining condition. The farm, owned by Haji Ali Sabri Kamaruddin, is located in a village with multi-floral surroundings. The team also observed several hives with plastic cups as pot replacement, but most of them were empty or closed with propolis.

Location 10 Meliponary in Kampung Lambor Kanan, Perak

Location 11: Meliponary in Kampung Ulu Rokan, Dangi, Gemencheh, Negeri Sembilan. Sampling began at 11:18 a.m., at 28°C under sunny weather. The farm, owned by Mohd Ghazali Othman, is located in a fruit farm with a multi-floral surrounding.

Location 11 Meliponary in Kampung Ulu Rokan, Dangi, Gemencheh, Negeri Sembilan

Location 12: Meliponary in Kampung Legung Ulu, Kota, Negeri Sembilan. Sampling began at 12:49 p.m., at 29°C and in sunny weather. The farm, owned by Mohd Zaki Haji Lasim, is located in a resident compound with a multiflora area. The hives were placed inside a locked hut covered with a wired fence.

Location 12 Meliponary in Kampung Legung Ulu, Kota, Negeri Sembilan

Location 13: Meliponary in Kampung Tanjung Ipoh, Negeri Sembilan. Sampling began at 5:17 p.m., at 30°C and in cloudy condition. The farm, owned by Azril Dino Abd Malik, is located in a suburban area, close to heavy traffic with multi-floral surroundings. The hives are equipped with thin topping boxes or supers to enhance pot formation and increase honey production.

Location 13 Meliponary in Kampung Tanjung Ipoh, Negeri Sembilan

Location 14: Meliponary in Kampung Lubuk Kulit, Kuala Lipis, Pahang. Sampling began at 1:27 p.m., at 32°C and in cloudy condition. The farm, owned by Zaimi Mohamed, is located in a remote secondary forest with a multi-floral surrounding. The farm has more than 100 colonies with various *H. itama* to several numbers of dipterocarp dependent species.

Location 14 Meliponary in Kampung Lubuk Kulit, Kuala Lipis, Pahang

Location 15: Meliponary in Felda Gunung Besout, Perak. Sampling began at 11:15 a.m., at 29°C and in sunny weather. The farm, owned by Azlin Mohd Hamim, is located in a resident compound with multi-floral surrounding fruit trees such as durian, mangosteen and rambutan and was close to an oil palm plantation.

Location 15 Meliponary in Felda Gunung Besout, Perak

Location 16: Meliponary in Lumut, Perak. Sampling began at 2:36 p.m., at 33°C and in cloudy condition. The farm, owned by Zakaria Abu Hassan, is located on the hillside of a secondary forest, with 80% of the area surrounded by Acacia plant species. The farm is situated near the beach, with exposure to hot air from the sea.

Location 16 Meliponary in Lumut, Perak

Location 17: Meliponary in Kemaman, Terengganu. Sampling began at 8:51 a.m., at 26°C and in cloudy condition. The farm owned by Nor Riah Mohamood is located in a village area with multi-floral surroundings near the beach, with exposure to hot air from the sea.

Location 17 Meliponary in Kemaman, Terengganu

Location 18: Meliponary in Felda Kerteh 2, Paka, Terengganu. Sampling began at 1:28 p.m., at 31°C and in cloudy condition. The farm, owned by Zaidi Anuar Kamarulzaman, is located in a village with multi-floral surroundings near the beach, exposed to hot air from the sea.

Location 18 Meliponary in Felda Kerteh 2, Paka, Terengganu

Location 19: Meliponary in Hutan Lipur Sekayu, Terengganu. Sampling began at 5:32 p.m., at 29°C and in rainy conditions. The farm, owned by Mohd Nurhafis Jaafr from the Department of Agriculture, is surrounded by a tropical forest close to Sekayu waterfall with a multi-floral environment.

Location 19 Location 19:

Location 20: Meliponary in Kampung Gemuruh, Kuala Nerus, Terengganu. Sampling began at 11:.00 a.m., at 33°C and in cloudy condition. The farm, owned by Mohd Nurazim Jaafar, is located in a village with multi-floral surroundings.

Location 20 Meliponary in Kampung Gemuruh, Kuala Nerus, Terengganu

Location 21: Meliponary in Kluang, Johor. Sampling began at 11:27 a.m., at 31°C in sunny weather. The farm, owned by Azizul Katan, is located in a durian farm with a multi-floral surrounding. The farm set with a hybrid concept, where the stingless bee and *Apis mellifera* hives were kept in the same. A wild *Apis cerana* hives found under several empty stingless bee topping roofs.

Location 21 Meliponary in Kluang, Johor

Location 22: Meliponary in Kampung Seri Dagang, Segamat, Johor. Sampling began at 3:36 p. m., at 33°C and in cloudy conditions. The farm, owned by Ranto Hidayat Syafii, is located in a rubber plantation with a multi-floral surrounding.

Location 22 Meliponary in Kampung Seri Dagang, Segamat, Johor

Location 23: Meliponary in Kampung Gemereh, Segamat, Johor. Sampling began at 5:46 p. m., at 31°C under sunny weather. The farm, owned by Rosli Ali, is located in a village with dominant and multi-floral resources.

Location 23 Meliponary in Kampung Gemereh, Segamat, Johor

Location 24: Meliponary in Kampung Sungai Terap, Tanjung Karang, Selangor. Sampling began at 11:09 a. m., at 30°C, under sunny weather. The farm, owned by Haji Misbah Haji Yusof, is located on a coconut farm, next to an oil palm plantation with a multi-floral surrounding. The farm is also situated near to the beach, with exposure to hot air from the sea.

Location 24 Meliponary in Kampung Sungai Terap, Tanjung Karang, Selangor

Location 25: Meliponary in Batu 17, Dusun Tua, Hulu Langat, Selangor. Sampling began at 11:00 a. m., at 30°C, under sunny weather. The farm, owned by Faridah Khalid, is located in a resident compound with multi-floral surrounding and fruit trees.

Location 25 Meliponary in Batu 17, Dusun Tua, Hulu Langat, Selangor

Location 26: Meliponary in Felda Palong 1, Gemas, Negeri Sembilan. Sampling began at 11:12 a. m., at 30°C and in cloudy condition. The farm, owned by Abu Sofian Ruslan, is located in a resident compound with multi-floral surrounding fruit trees and an oil palm plantation.

Location 26 Meliponary in Felda Palong 1, Gemas, Negeri Sembilan.

Location 27: Meliponary in Felda Serting 4, Bandar Seri Jempol, Negeri Sembilan. Sampling began at 1:09 p. m., at 31°C and in cloudy condition. The farm, owned by Ehwani Manap, is located in a resident compound with multi-floral surrounding fruit trees and an oil palm plantation.

Location 27 Meliponary in Felda Serting 4, Bandar Seri Jempol, Negeri Sembilan

Location 28: Meliponary in Kampung Butut, Kuala Berang, Terengganu. Sampling began at 3:16 p. m., at 27°C and in cloudy condition. The farm, owned by Mohd Alias Mamat, is located in a village with a multi-floral surrounding.

Location 28 Meliponary in Kampung Butut, Kuala Berang, Terengganu

Location 29: Meliponary in Kampung Machang, Ajil, Terengganu. Sampling began at 12:18 p. m., at 30°C and in cloudy condition. The farm, owned by Suhaimi Salleh, is located in a village area with Tapioca species as a dominant source.

Location 29 Meliponary in Kampung Machang, Ajil, Terengganu

Location 30: Meliponary in D'Bullah Garden, Serendah, Selangor. Sampling began at 9:56 a. m., at 27°C and in cloudy condition. The farm, owned by Hasbullah Hamzah, is located in a fruit farm next to Fraser's Hill's main road with the multi-floral surrounding.

Location 30 Meliponary in D'Bullah Garden, Serendah, Selangor

Location 31: Meliponary in Tanjung Malim, Perak. Sampling began at 1:03 p. m., at 29°C and in cloudy condition. The farm, owned by Mr Hendra Ja, is located in the suburban area, with fruit trees and multi-floral surroundings.

Location 31 Meliponary in Tanjung Malim Perak,

Location 32: ***Apis cerana*** **farm in Seri Kembangan, Selangor** Sampling began at 10:51 a. m., at 31°C and in cloudy condition. The farm, owned by Ahmad Nizar Masandi, is located in the Acacia Forest in an urban area.

Location 32 Apis cerana farm in Seri Kembangan, Selangor

Location 33: Meliponary in Selangor Fruit Valley, Selangor. Sampling began at 10:06 a. m., at 27°C and in cloudy condition. The farm, owned by Sazali Pariuddin, is located in a designated fruit valley surrounded by Acacia trees.

Location 33 Meliponary in Selangor Fruit Valley, Selangor

Location 34: Meliponary in Jelebu, Negeri Sembilan. Sampling began at 11:05 a. m., at 29°C and was in cloudy condition. The farm, owned by Jeffry Iskandar Foo, is located in a village with multi-floral surroundings.

Location 34 Meliponary in Jelebu, Negeri Sembilan

Location 35: Meliponary in Kuala Kangsar, Perak. Sampling began at 9:24 a. m., at 27°C, under sunny weather. The farm, owned by Noor Hanim Aziz, is located in a village with various fruit trees and multi-floral surroundings.

Location 35 Meliponary in Kuala Kangsar, Perak

Location 36: Meliponary in Lenggong, Perak. Sampling began at 11:59 a. m., at 29°C, under sunny weather. The farm, owned by Muhamad Ariff Amri, is located in a suburban area with a multi-floral surrounding.

Location 36 Meliponary in Lenggong, Perak

Location 37: Meliponary in Batangsi, Selangor. Sampling began at 9:30 a. m., at 26°C and was in cloudy condition. The farm, owned by Tuan Aminaddin, is located in a fruit farm with dominant and multi-floral resources.

Location 37 Meliponary in Batangsi, Selangor

Location 38: Meliponary in Kampung Pulau Jerejak, Bukit Lada, Chenor, Pahang. Sampling began at 10:20 a. m., at 27°C under sunny weather. The farm, owned by Mohd Nor Firman, is located in a village with various fruit trees and multi-floral surroundings. The farm is also situated near to swallow birdhouse.

Location 38 Meliponary in Kampung Pulau Jerejak, Bukit Lada, Chenor, Pahang

Location 39: Meliponary in Jerantut, Pahang. Sampling began at 3:45 p. m., at 29°C and in cloudy conditions. The farm, owned by Syed Norhaimi Syed Zin, is located in Salak farm, near oil palm plantation and multi-floral surroundings.

Location 39 Meliponary in Jerantut, Pahang

Location 40: Meliponary in Kampung Sawah Dato, Mersing, Johor. Sampling began at 10:02 a. m., at 25°C and under sunny weather. The farm, owned by Md Safari Sulaiman, is located in a coconut farm, next to an oil palm plantation and multi-floral surrounding. The farm is also situated near to the beach, with exposure to hot air from the sea.

Location 40 Meliponary in Kampung Sawah Dato, Mersing, Johor

Location 41: ***Apis mellifera*** **farm in Mersing Bee Farm, Johor**. Sampling began at 2:37 p. m., at 28°C and in cloudy condition. The farm, owned by Datuk Saidon, has 1,000 hives located in an oil palm plantation with a multi-floral surrounding.

Location 41 Apis mellifera farm in Mersing Bee Farm, Johor

Location 42: Meliponary in Langkawi, Kedah. Sampling began at 9:31 a. m., at 29°C, under sunny weather. The farm, owned by Mohd Nahizam, is located in a museum vicinity, near the secondary forest and village area with the multi-floral surrounding.

Location 42 Meliponary in Langkawi, Kedah

Location 43: Meliponary in Perlis. Sampling began at 3:09 p. m., at 33°C and in cloudy conditions. The farm, owned by Roshita Hussin, is located in a village area with fruit trees and multi-floral surroundings.

Location 43 Meliponary in Perlis

Location 44: Meliponary in Baling, Kedah. Sampling began at 9:32 a. m., at 25°C and in cloudy conditions. The farm, owned by Radiga Ramanan, is located in a rubber plantation with a multi-floral surrounding. The farm is located near a metal scrap workshop.

Location 44 Meliponary in Baling, Kedah

Location 45: Meliponary in Bayan Lepas, Pulau Pinang. Sampling began at 10:00 a. m., at 28°C under sunny weather. The farm, owned by Raymond Sim, is located in a durian farm and multi-floral surrounding.

Location 45 Meliponary in Bayan Lepas, Pulau Pinang

Location 46: Meliponary in Kampung Bintang, Batu Pahat, Johor. Sampling began at 8:17 a. m., at 24°C and in cloudy condition. The farm, owned by Shamsul Bahri Ahmad, is located in a coconut, cocoa and coffee hybrid farm with a multi-floral surrounding.

Location 46 Meliponary in Kampung Bintang, Batu Pahat, Johor

Location 47: Meliponary in Parit Sekoyan, Kesang Laut, Muar, Johor. Sampling began at 12:27 p. m., at 31°C, under sunny weather. The farm, owned by Firdaus Hafiz, is located in a coconut farm with a multi-floral surrounding.

Location 47 Meliponary in Parit Sekoyan, Kesang Laut, Muar, Johor

Location 48: Meliponary in Bagan Lalang, Sepang, Selangor. Sampling began at 9:33 a. m., at 26°C and in cloudy condition. The farm, owned by Roy Khoo, is located in a mangrove swamp by the beach. The farm is exposed to hot air almost throughout the year. The honey from this area's significant characteristics is a strong, pungent, mouldy smell with a bitter and sour taste. Some people might find it almost inedible, but 90% of the customers reported using honey as a cough remedy.

Location 48 Meliponary in Bagan Lalang, Sepang, Selangor

Location 49: Meliponary in Kampung Semarak, Pasir Puteh, Kelantan. Sampling began at 10:33 a.m., at 27°C and in cloudy condition. The farm, owned by Mohd Faizal, is located in a mangrove swamp area (80% dominant) with a multi-floral surrounding.

Location 49 Meliponary in Kampung Semarak, Pasir Puteh, Kelantan

Location 50: Meliponary in Ketereh, Kelantan. Sampling began at 1:37 p.m., at 27°C and in cloudy condition. The farm, owned by Kamarulzaman Nasir, is located in a village area next to a rubber plantation and fruit farm with a multi-floral surrounding.

Location 50 Meliponary in Ketereh, Kelantan

Location 51: Meliponary in Kem Desa Pahlawan, Kok Lanas, Kelantan. Sampling began at 9:34 a.m., at 25°C and in cloudy condition. The farm, owned by Rosman Mamat, is located in a military base with Acacia surrounding it. A significant characteristic of the honey in this area is its thick structure and sweet taste.

Location 51 Meliponary in Kem Desa Pahlawan, Kok Lanas, Kelantan

Location 52: Meliponary in Permatang Badak, Kuantan, Pahang. Sampling began at 1:46 p. m., at 32°C under sunny weather. The farm, owned by Wan Ahmad Shamsuddin, is located in Acacia forest with a multi-floral surrounding.

Location 52 Meliponary in Permatang Badak, Kuantan, Pahang

Results and Discussion

The sampling activities that began on July 12, 2019, came to an end on August 3, 2020. All honey samples collected showed a signature feature of stingless bee honey, with high humidity and sour taste. On average, H. itama honey's humidity was around 28.2 % and ranging from 18.5% (lowest) to 30.5% (highest). The sugar contents are lower than previously reported in the available standard, where the majority of samples from all locations showed less than 50% total sugars (unpublished data). The flexibility is the main characteristic of *the H. itama colony, as proven by its high survival rate in various environmental surroundings*, making *H. itama* the best choice for meliponary farming.

REFERENCES

1. Ab Hamid, S., Salleh, M. S., Thevan, K., & Hashim, N. A. 2016. Distribution and Morphometrical Variations of Stingless Bees (Apidae: Meliponini) In Urban and Forest Areas of Penang Island, Malaysia. Journal of Tropical Resources & Sustainable Science 4: 1-5

2. Hrncir, M., Jarau, S., & Barth, F. G., 2016. Stingless bees (Meliponini): senses and behaviour. Journal of Comparative Physiology A, 202:597–601

3. Azizi, M. G., Priawandiputra, W., & Raffiuddin, R. 2020. Morphological identification of stingless bees from Belitung. IOP Conference Series: Earth and Environmental Science 457; doi:10.1088/1755-1315/457/1/01/012011

Chapter 17

17.0 Advances in Beescape for Meliponines

17.1. Plantings in 4 strata levels

Figure 100 Planting and vegetation strata levels

Looking closely at the vegetation of a vibrant Meliponary, one will realize that there are mainly four strata levels in the Meliponines foraging realm. More likely, the stories would appear as follows:

Level 1 – Major tree floral canopy or feature palm spadices

Level 2 – Saplings and Structural shrubs & climbers or ramblers

Level 3 – Shrubbery or hedges and low flowering plants

Level 4 - Flowering ground covers & ground creepers

Scrutinizing with a keener eye, one may see more levels. Still, to plan the provisions necessary for a basic Meliponary, these levels or groupings are adequate for creating ample foraging sources.

17.1.1 Resource Partitioning

The levels are beneficial when the farm is rampant with predatory insects. It provides for natural resource partitioning and choices of resources to avoid competition among foraging enemies. Also, for inter-species, especially when one has a few different meliponine species of other preferences and diverse flight ranges.

17.2. Communication with the bees in Beescape planning

Figure 101 Example Beescape compositions

One would design and plan a recreational park, locating the main entrance, walkways, rubbish bins, benches, gazebos or rest huts, playground, toilets, par course etc., these elements were placed in a manner to communicate with the park visitor without the spoken word. Nonverbal suggestions like

- Archway entrance implies "*enter here*";
- Steppingstones or an elaborate walkway says, "*walk this way and don't step on the grass*";
- Rubbish bins strategically placed tells you, "*Throw your trash here*";
- Garden benches say, "*rest here if your legs are weary*";
- Gazebos and rest huts indicate "*temporarily shelter here in extreme weather like rain or direct hot sun*";
- Parcourse "*exercise here*" and children's playground "*let your children play here*" and subsequently, "*recreational activities*",
- some water body may be incorporated;
- vast areas of lawn for picnics and family sports hobbies.

So, all these elements are planned and placed strategically to allow a controlled movement or circulation of space without the need to churn out verbal instructions.

By this same token, one may have "nonverbal communication" with the bees without spoken or verbal instructions, e.g.

- Have their favourite climbers at perimeter fences or trellises to limit their range, especially if the area beyond the fence is not entirely pesticide-free.
- Ravenala Palms, Strelitzias or Heliconias can serve as great windbreakers while providing excellent blooms for nectar foraging.
- Include plants that have racemes and downward floral showers. Avoid upward flower heads because they are susceptible to aerial predating insectivorous birds.
- Hedges that would limit their range of foraging with thick bushy shrubs like *Baphia nitida* and the likes. Do not go beyond where dangers lurk.
- Groundcovers that flowers sufficiently that would tell the bees to stay low and avoid competing enemies and predators in flight.
- Waterbodies or sponge water troughs or basins with clean water to avoid foraging unclean waste fluids like animal urine or dirty swales.
- Salts and minerals in the form of dry salted fish. Fresh beach pebbles or brine dipped used teabags within the Meliponary area.
- Strategically timed feeding stations for fruit peels and rind or dry feeds like broken corn, crushed peas or powdered nuts or even cassava flour.

17.3. Meliponary Topsoil, Relative humidity and Groundcovers

Groundcover or ground cover is any plant that grows over an area of ground. Groundcover protects the topsoil from erosion and drought. The ground cover forms the layer of vegetation below the shrub layer known as the herbaceous layer in an ecosystem.

Ground covers are perennial plants or, sometimes, low-growing shrubs that form a stable, solid mat of growth, replacing either lawn or garden beds. It is best not to mix too many kinds of ground covers into the same area, for they will spread and become intermixed in a way that is hard to handle. Soil bed borders are an excellent way to maintain them. They regulate ideal relative humidity for Meliponines.

Examples of appropriate low-maintenance ground covers for your Meliponary.

1. Portulacas or Purslane (Japanese rose)
2. Mimosa (forget me not)
3. *Ipomoea Batatas* (Sweet potato)

4. Convolvulaceae creepers (Water spinach, Kangkung or morning glory)
5. Shade tolerant – *Hemigraphis*, *Rhoeo*, Bleeding Hearts
6. Low flowering tuft like shrubbery *Vinca*, *Zephyranthes*, *Cuphea*

17.4 Decaying n decomposing organic matter

Poor or lack of groundcover vegetation can harbour fungus and bacteria in the decomposing litter. Prone to drought and pest beetles, flies and other predatory elements like spiders and lizards. Comparatively, the one with vegetative ground cover appears more aesthetically pleasing

Figure 102 Poor or lack of groundcover vegetation can harbour fungus and bacteria in the decomposing litter.

17.5. Keeping up honey grade.

Beescape and farm management per FAO guidelines.

- Box designs for honey production and pollination according to species adaptation.
- Eduction methodology and swarm trap incorporate proper preparation of bait.
- Pollination of cash crops and permaculture was scrutinized.
- Colony biology and gender or queen development techniques incorporated.
- Strategies in entrance tube development and colony health assessment applied.
- Colony Drifting is prevented by hive density and placement control with weather adaptation.
- Sustainability and conservation efforts are well implemented.
- Seasonal vegetation incrementation and resource partitioning review
- Preparations for extreme weather conditions.
- Harvesting schedules and post-harvest storage are well planned.
- Products and derivatives extraction, preparations and inventory storage at correct requirements

17.6 Windbreakers n perimeter planting.

Ravenella vs *Baphia nitida*.

Figure 103 Perimeter planting to restrict bee flight range

17.7. Managing swarms in a Meliponary

In Managing swarms, the hive box designs and finishes become very critical. To avoid drifting and intermingling swarms from causing colony battling, a good way is to colour the hive boxes of different shades and tones. See the appendix of swarm types.

17.8. Foraging peculiarities

Figure 104 Foraging peculiarities

What are these bees doing?

There are many possibilities.

1. There may be "honeydew" droplets of aphids or treehoppers.

2. There may be an early stage or beginning of a male congregation.

3. Processing or suctioning of extrafloral nectar from the seepage of other leaves.

4. Maybe collecting pollen blown by the wind.

Nectar is very watery (about 65% moisture). It thickens into honey, and in my opinion, the behaviour of meliponines is like removing liquid in its mouth and then being sucked again. It is done by bees repeatedly on the leaf surface under sunlight to reduce the nectar's water content into honey.

* Personal assumptions

This explanation is meant earlier. I agree with this, but not merely because the nectar is runny. Nectar has 65% water, and the processing is carried out in conjunction from one mouth to another. The photo shows only three numbers of bees. So, I listed the other possibilities and a question posted as follows:

Q: I have documented the behaviour of stingless bees crowding in the sand. What are they looking for, sir?

A: Maybe there are unclean remnants in the sand, and sometimes they take "loam" for the geopropolis mixture. Loam or clayey soil refers to ingredients in soil constituencies.

https://images.app.goo.gl/5Vp9v8ZM3XggrtgG9

They sometimes collect pollen scattered in the leaves only that the pollen is too microscopic to be seen with the naked eye.

Figure 105 collect pollen scattered in the leaves

There were times when they licked 'Mucilage'. Mucilage is a mucus substance in plants such as aloe vera, cactus, okra, and banana tree trunks. Mucilage seepage from some vegetation, and the leaves or droplets from adjacent plants are on the leaves' surface. The strong wind then sprayed it on deck.

Figure 106 Licking up mucilage from nestmate.

The workers here at the end of this Bougainvillea twig seems to have something mucilaginous on its thorax, which the other bee is licking. It is uncertain what bees do with mucilage or that the worker's gummy looking substance is mucilage. It appears like very viscous nectar, or sap, or something like resin.

It just goes to show that there is still much to be discovered about bee biology.

17.9 Landscape jargon in the ecological context

- land use

How humans use a particular area of land (e.g. urban, agricultural, industrial etc.)

- land-use change

The process of changing the land use type of a particular area of land (e.g. from agricultural to residential) significantly when the land has not previously changed from its original use or has been under the actual use for a very long time.

- landrace

The varieties of crops developed in an area and thus are tolerant of the area's environmental stresses can produce medium to high and consistent yields with low inputs in that specific area.

- landscape

A particular geographical region's fundamental attributes include its land cover and land use patterns, biological and physical characteristics.

- landscape context

The position and shape of patches in a fragmented landscape concerning the landscape as a whole (Johnson et al. 2007)

- landscape ecology

The study of landscapes concerning the ecology of their biological populations.

- landscape indicator

Any element (quantitative or qualitative) of a landscape used for ongoing assessment and measurement of the landscape's overall values.

- landscape level

Refers to studies or activities carried out over an area within the range of several to tens of kilometres, encompassing different habitat types etc.

- landscape physiognomy

The collective features associated with the physical layout of elements within the landscape (Dunning et al. 1992)

- landscape structure

The collective attributes (e.g. size, shape, position, composition etc.) of the ecosystems/habitats make up a landscape.

- landscape variable

A variable is used to describe a single patch of habitat and the spatial distribution of that patch in space or within the whole landscape structure.

- landscape-scale study

Type of ecological research aimed at determining how landscape structure affects species distribution and abundance, which is done by studying several non-overlapping landscapes of different forms rather than one ecosystem or area.

- logging impact

Any impact on a natural system that is caused by the process of logging or the subsequent absence of trees post-logging; effects include changes in biodiversity levels, vegetation structure, hydrology and habitat quality.

- logging-induced mortality

Mortality in bee nests is caused by the harvesting of nest trees, which either destroy nests or allows greater predator access or increased environmental exposure. (Eltz et al. 2003)

Bibliography

Allsopp, M., Lange, W. J., & Veldtman, R. (2008). *Valuing Insect Pollination Services with Cost of Replacement.* Journal pone.

Amano, K. (2004). *ATTEMPTS TO INTRODUCE STINGLESS BEES FOR THE POLLINATION OF CROPS UNDER GREENHOUSE CONDITIONS IN JAPAN.* National Institute of Livestock and Grassland Science Japan.

Anonymous. (2008). *Tools for Conservation and Use of Pollination Services - Initial SURVEY OF GOOD POLLINATION PRACTICES.* FAO.

Anonymous. (n.d.). *Bees and their role in forest livelihoods - THE VALUE OF BEES FOR CROP POLLINATION.* FAO.

Anonymous. (n.d.). *Pollination Manual - Reproductive Biology and Pollination in Rainforest Trees.* Reproductive Biology and Pollination in Rainforest Trees.

Ascher, J., & Pickering. (28 Feb 2018). *Discover Life bee species guide and world checklist,* Retrieved from Wallacetrigona: http://www.discoverlife.org/mp/20q?search=Wallacetrigona&guide=1&flags=HAS:subgenus :

Azmi, W. A., Seng, C. T., & Solihin, N. S. (2016). *POLLINATION EFFICIENCY OF THE STINGLESS BEE, Heterotrigona itama (HYMENOPTERA: APIDAE) ON CHILI (Capsicum annuum) IN GREENHOUSE.* J. Trop. Plant Physiol.

Barfod, A. S., Hagen, M., & Borchsenius, F. (2011). *Twenty-five years of progress in understanding pollination mechanisms in palms (Arecaceae).* Annals of Botany, Volume 108, Issue 8,

Birt, M. (n.d.). *The Trigona Species (Stingless Bees). Their role in higher plant pollination.*

Boongird, M. &. (2004). A New Species of Trigona from Peninsular Thailand (Hymenoptera: Apidae: Meliponini).

Boontop, Y., Malaipan, S., & Wiwatwittaya, K. C. (2008). Diversity of Stingless Bees (Apidae: Meliponini) in Thong Pha Phum District, Kanchanaburi Province, Thailand. *Kasetsart J. (Nat. Sci.) 42*, 444 - 456.

Cervancia, C. (2017). *Asian Pollination Initiative.* UPLB.

Chambo, E., Garcia, R., de Oliverira, N. T., & Doarte-Junior, J. (2011). Honey bee visitation to sunflower effects on pollination and plant genotype. *sci, Aqric. (Piracic).*

Chanchao, C. (2009). ANTIMICROBIAL ACTIVITY BY TRIGONA LAEVICEPS (STINGLESS BEE) HONEY FROM THAILAND. *Pak J Med Sci April - June 2009 (Part-II) Vol. 25 No. 3*, 364-369.

Chinajariyawong, A., & Saiboon, S. (n.d.). Culture of the stingless bee, Trigona laeviceps Smith (Hymenoptera: Apidae). *Institute of Agricultural Technology, Walailak University.*

Chuttong, B., & Burgett, M. (2016). Biometric studies of the stingless bee Tetragonula laeviceps species complex (Apidae: Meliponini) in SE Asia (Thailand). *1st International Meliponine Conference.*

Ciar, R. R., & et al. (n.d.). FORAGING BEHAVIOR OF STINGLESS BEES (Tetragonula biroi Friese): DISTANCE, DIRECTION AND HEIGHT OF PREFERRED FOOD SOURCE. *UPLB BEE Program.*

Cimu, I. (1960). Results of bee pollination of Sunflowers. *Apicultura. 33(1) :* , 18-20.

Cockerell, T. D. (1919). The Social Bees of The Philippines Islands. The *University of Colorado.*

COHN, D. (2004). Xunan Kab, The Stingless Bees of the Yucatan: Preserving meliponiculture in Mayan communities. *American Bee Journal.*

Coolborn, A. F., Bolatito, B., & Clement, A. F. (2012). Study of Acute and Sub Chronic Toxicity of Spathodea campanulata P Beav Leaf. *IPCBEE vol.41.*

CORTOPASSI-LAURINO, M., & al, e. (2006). Global meliponiculture: challenges and opportunities. *Apidologie 37*, 275–292.

COUVILLON, M. J., WENSELEERS, T., IMPERATRIZ-FONSECA, V. L., NOGUEIRA-NETO, P., & RATNIEKS, &. F. (2008). A comparative study in stingless bees (Meliponini) demonstrates that nest entrance size predicts traffic and defensivity. *J. EVOL. BIOL. 21*, 194–201.

DANARADDI, C. S., VIRAKTAMATH, S., & BHAT, K. B. (2009). Nesting habits and nest structure of stingless bee, Trigona iridipennis Smith at Dharwad, Karnataka. *Karnataka J. Agric. Sci., 22(2):*, (310-313).

Eltz, T., & al., e. (2003). Nesting and nest trees of stingless bees (Apidae; Meliponini) in lowland dipterocarp forest in Sabah, Malaysia, with implications for forest management. *Forest Ecology and Management 172.*

Engel, M. S., & Rasmussen, C. (2017). A new subgenus of Heterotrigona from New Guinea (Hymenoptera: Apidae). *Journal of Melittology.*

Free, J. B. (n.d.). *POLLINATION IN THE TROPICS.* CMB, UK.

Gallai, N., & Vaissière, B. E. (2009). *Guidelines for the economic valuation of pollination services at a national scale.* FAO.

Gallaia, N., Sallesc, J.-M., Setteled, J., & Vaissièrea, B. E. (2009). *Economic valuation of the vulnerability of world agriculture confronted with pollinator decline.* Science Direct.

Heard, T. (2016). The Australian Native Bee Book, *ISBN 978-0-646-93997.*

Heard, T. A. (1999). *THE ROLE OF STINGLESS BEES IN CROP POLLINATION.* Annu. Rev. Entomol. . 44:183–206.

HENDERSON, A. (1986). *A Review of Pollination Studies in the Palmae.* The Botanical Review.

Imperatriz-Fonseca, V. L. (n.d.). *Best management practices in Agriculture for sustainable use and conservation of pollinators.* FAO.

Inoue, T., Salmah, S., Abbas, I., & Yusuf, E. (1985). 1985. Foraging Behavior of Individual Workers and Foraging Dynamics of Colonies of Three Sumatran Stingless bees. *Res. Popul. Ecol. 27:*, 373-392.

Jaapar, M. F., & Jajuli, R. (2014). *Stingless bees: Utilization of stingless bees pollinating rockmelon under the greenhouse.* MARDI.

Jalil, A. (2014). *Beescape for Meliponines.* Singapore: Partridge Publishing.

Jalil, A. (2016). *Meliponine Identifier Pictorial Guide.* Akademi Kelulut Malaysia.

Jalil, A. H., & Shuib, I. (2014). *Beescape for Meliponines - Conservation of Indo-Malayan Stingless Bees.* Partridge Singapore.

Jalil, A., & Roubik(ed), D. (2016). *Handbook of Meliponiculture.* Akademi Kelulut Malaysia.

Jalil, A., & Roubik(ed), D. (2018). Handbook of Meliponiculture Vol 2. *Akademi Kelulut malaysia.*

KASINA, J. M., MBURU, J., KRAEMER, M., & HOLM-MUELLER, K. (2009). *Economic Benefit of Crop Pollination by Bees: A Case of Kakamega Small-Holder Farming in Western Kenya.* J. Econ. Entomol.

KishanTej, M., Srinivasan, M. R., & Thakur, R. K. (2017). *Stingless bees for applied pollination in cucumber greenhouses of South India.* AKM-IMS3C.

Klein, A.-M., & al., e. (2007). *Importance of pollinators in changing landscapes for world crops.* Proc. R. Soc. B.

Luttrell, R. B. (2016). *African-tulip-tree-ebook.* Retrieved from www.bobthebeeman.com.au: http://bobthebeeman.com.au/african-tulip-tree-ebook.asp

Luttrell, R. (n.d.). *bobthebeeman.com.au.* Retrieved from explaining-the-Aussie-inpa-box: http://bobthebeeman.com.au/explaining-the-aussie-inpa-box.htm

MICHENER, C. (1974). The Social Behavior of the Bees: a Comparative Study. *Harvard University Press, Cambridge, MA.*

Michener, C. D. (2007). *The Bees of the World.* Baltimore: The Johns Hopkins University Press.

Michener, C. D. (2007). The Bees of the World 2nd ed. *The John Hopkins university press.*

Momose, K., & al., e. (1998). *Pollination Biology In A Lowland Dipterocarp Forest In Sarawak, Malaysia. I.Characteristics Of The Plant-Pollinator Community In A Lowland Dipterocarp Forest.* American Journal of Botany.

Moure, I. (1961). Preliminary Supra-specific classification of the old world Melliponinae Bees (Hymenoptera, Apoidea). *Studia Entomology. 4:* 181-242.

Nogueira-Neto, P. (1997). Life Creating Bees.

Norowi & et al. (2008). *Conservation and sustainable utilization of stingless bees for pollination.* MARDI.

Occhiuzzi, P. (2000). Stingless bees pollination greenhouse Capsicum, *Aussie Bee 13,15. Published by Australian Nature Bee Research Centre, North Richmond, NSW Australia.*

Raes, N., & et al. (2014). Historical distribution of Sundaland's Dipterocarp rainforests at Quaternary glacial maxima. *PNAS.*

Rahmanl, A., Das, P. K., Rajkumari, P., Saikia, J., & Sharmah, D. (2015). Stingless Bees (Hymenoptera:Apidae: Meliponini): Diversity and Distribution in India. *International Journal of Science and Research (IJSR)*, Volume 4, Issue 1.

Rasmussen. (2008). Catalogue of the Indo-Malayan/Australasian stingless bees (Hymenoptera: Apidae: Meliponini).

Rasmussen, C. (2008). Molecular Phylogeny Of Stingless Bees: Insights Into Divergence Times, Biogeography, And Nest Architecture Evolution (Hymenoptera: Apidae: Meliponini). The *University of Illinois at Urbana-Champaign.*

RASMUSSEN, C. (2013). Stingless bees (Hymenoptera: Apidae: Meliponini) of the Indian subcontinent: Diversity, taxonomy and current status of knowledge. *Zootaxa 3647 (3): 401–428.*

Rasmussen, C., & Camargo, J. M. (2008). Molecular phylogeny and the evolution of nest architecture and behaviour in Trigona s.s. (Hymenoptera: Apidae: Meliponini). *Apidologie 39.*

RASMUSSEN, C., THOMAS, J. C., & ENGEL3, M. S. (2017). A New Genus of Eastern Hemisphere Stingless Bees (Hymenoptera: Apidae), with a Key to the Supraspecific Groups of Indomalayan and Australasian Meliponini. *AMERICAN MUSEUM NOVITATES.*

Ricketts, T. H., & al., e. (2008). *Landscape effects on crop pollination services: are there general patterns?* Ecology Letters.

Roubik, D. (1995). Pollination of cultivated plants in the tropics, *FAO Agric. Serv. Bull. 118, Rome.*

Roubik, D. (2006). Stingless bee nesting biology. *Apidologie 37.*

Roubik, D. W. (1993). *Direct costs of forest reproduction, bee-cycling and the efficiency of pollination modes.* STRI Panama.

S.D., L., Zeilhofer, S., Blu¨ thgen, N., & Schmitt, T. (2010). Stingless Bees Use Terpenes as Olfactory Cues to Find Resin Sources. . *Chem. Senses 35:* , 603-611.

sakagami, S. F. (1975). Stingless Bees (Excl. Tetragonula) from Continental Southeast Asia in the Collection of Bernice P. Bishop Museum, Honolulu (Hymenoptera, Apidae). *Jour. Fac. Sci. Hokkaido Univ., Ser. VI, Zool. 20 (1),.*

Sakagami, S. F. (1978). Tetragonula Stingless Bees of the Continental Asia and Sri Lanka (Hymenoptera, Apidae). *Jour. Fac. Sci. Hokkaido Univ. Ser. VZ, 2001. 21(2).*

Sakagami, S. F., & Inoue, T. (1985). Taxonomic Notes on Three Bicolorous Tetragonula Stingless Bees in Southeast Asia. *Kontyii, Tokyo, 53 (1): 174-189.*

Sakagami, S. F., Inoue, T., & Salmah, S. (1990). Stingless Bees of Central Sumatra. *Sapporo; Hokkaido University Press.*

SANTOS, S. A., ROSELINO, A. C., & BEGO, L. R. (2008). *Pollination of Cucumber, Cucumis sativus L. (Cucurbitales: Cucurbitaceae), by the Stingless Bees Scaptotrigona aff. depilis*

Moure and Nannotrigona testaceicornis Lepeletier (Hymenoptera: Meliponini) in Greenhouses. Neotropical Entomology.

Santos, S. B., Roselino, A., Hrncir, M., & Bego, L. (2009). *Pollination of tomatoes by the stingless bee Melipona quadrifasciata and the honey bee Apis mellifera (Hymenoptera, Apidae).* GMR Journal.

Sawatthum, A., Tonwitoowat, R., & Umnuaysit, K. (2017). *THE ROLE OF STINGLESS BEE AND EUROPEAN HONEY BEE IN POLLINATION OF CONFECTIONERY SUNFLOWER.* Bangkok: AKM-IMS3C.

SCHWARZ, H. F. (1937). BORNEAN STINGLESS BEES OF THE GENUS TRIGONA. *Bulletin American Museum of Natural History.*

Schwarz, H. F. (1939). The Indo Malayan Species of Trigona. *Bulletin of AMNH.*

Sheffield, C. S., Ngo, H. T., & Azzu, N. (2016). *A Manual On Apple Pollination.* Fao.

Slaa, E. J., Chavez, L. A., Malagodi-Braga, K. S., & Hofstede, F. E. (2006). *Stingless bees in applied pollination: practice and perspectives.* Apidologie 37.

SOMMEIJER, M. (1999). Beekeeping with stingless bees: a new type of hive. *Bee World 80(2):* 70-79.

Tangmitcharoen, S., Takaso, T., Siripatanadilox, S., Tasen, W., & Owens, J. (2006). *The behaviour of major insect pollinators of teak (Tectona grandis L. f.): A comparison of clonal seed orchard versus wild trees.* Forest Ecology and Management.

Thornton, J. (2000). Environmental Impacts of Polyvinyl Chloride (PVC) Building Materials. *Columbia University's Earth Institute.*

Thummajitsakul, S., Klinbunga, S., Smith, D., & Sittipraneed, S. (2008). Genetic diversity and population structure of Trigona pagdeni Schwarz in Thailand. *Apidologie 39*, 446–455.

TRAVERO, J. T., NUEZ, I. L., & SAGLIBA, W. S. (2012). *Pollination of Apis mellifera and Trigona biroi on the.* IJERD.

Vasanthakumar, S., Srinivasan, M. R., & Thakur, R. K. (2017). *Pollination behaviour of stingless bees Tetragonula iridipennis Smith on mango inflorescence in south India.* AKM- IMS3C.

Vit, P., Roubik, D. W., & Pedro-(eds), S. M. (2013). Pot Honey - A Legacy of Stingless Bees. *Springer.*

List of Figures

Figure 1 Typical portable box hive for pollination 14
Figure 2 Durable 3-tiered box hive 16
Figure 3 Meliponines & Flower distinctiveness 29
Figure 4 Cucurbit Pollination 29
Figure 5 Honeydew melon trellis 30
Figure 6 Pesticide usage predicament 31
Figure 7 Revamping and redesign of greenhouse features 32
Figure 8 Portable Hive boxes and Field Hive Racks 33
Figure 9 Geniotrigona thoracica gender and caste 80
Figure 10 New arrivals in transit area 80
Figure 11 Placement under some natural shade and close to a body of water. Once placed in the farm proper, the base of the trunks was smeared with used motor oil or grease to prevent crawling insects like borers, ants, termites and even carpenter bees. This will preserve the wood as well. 80
Figure 12 Various foraging sources provided in the Meliponary 81
Figure 13 foraging minerals at a browning patch of turf 82
Figure 14 Nest entrance tubes pf Geniotrigona nests collected in Borneo. 82
Figure 15 Nest Structure of a feral hive in a tree hollow. 83
Figure 16 A. Horizontal spiralled (anticlockwise) layered brood cells. B. The first thing a colony does is build a thick dark involucrum. C. Initial array of pillars to hold storage pots. 84
Figure 17 Examples of weak colonies taken over by a smaller species Lepidotrigona sp. 86
Figure 18 Comparison of Geniotrigona spp. with Wallacetrigona sp. - one of the reasons G. incisa is now placed in a new genus on its own. i.e., Wallacetrigona. 87
Figure 19 Left: Natural honey storage pots unruly with 500-rupiah coin for scale indication, Middle: Actual thumb-size artificially moulded pot. Right: Arranged to occupy maximum space and for ease of expected harvesting. 88
Figure 20 A topping box laid with artificial plastic pots. 89
Figure 21 First trials of wax propolis moulded storage pots randomly placed to observe whether the bees accept them. What was placed was a replicated bottom half of the storage pots D, Pots (circled in red) being built up from the bottom half that was provided earlier 91
Figure 22 Progress of moulded pot trials for a medium-sized species. 92
Figure 23 A. from left: Dr. Ibrahim Shuib, AHJ, Radzi Alias and Yusnizar Ibrahim (Photographer); B. & C. Heterotrigona itama entrance activities, D. Discoloured propolis possibly due to fungus; E. & F. Hives under shade nets; G. Hives destroyed by Megachilid bees after massive cerumen pillage. 94
Figure 24 A. & B. Healthy H. itama brood; C. Tetrigona apicalis entrance hole at box hive; D. Tetragonula laeviceps box hive 95
Figure 25 A. Meliponary sign boards sponsored by Govt. Agency; B. Understorey of the coconut farm; C - G. Various sections of the farm with creative roofing. Yellow painted zinc roof & aluminium lined polycarbonate roofs with new dimension (19" x 32"). Overly large size dimensions to ensure sufficient coverage from rain and sun; H. Mr. Radzi in the new Meliponary. 96
Figure 26 A. & D. Harvesting equipment (Honey suction pumps); B. Full Harvesting gear; C. H. itama honey pots; E. & F. pollen pots; G. Tetragonula laeviceps storage pots and cells; H. & J. Extraordinary Large honey pots of H. itama. 97
Figure 27 The Meliponary in a coconut farm (pandan variety) 98
Figure 28 Brown Sea cucumber (or "deep-water redfish"), Actinopyga echinites. used in the concoctions. 98
Figure 29 Peperomia pellucida, Pepper Elder, locally known as 'sinaw-sinaw' in the Philippines (literally 'shining') 98

Figure 30 Sea cucumber maceration in Honey *99*
Figure 31 A. Harvested Pollen (pot-pollen); B. Selected herbs for cosmeceuticals; C. Packaged products; D. Assorted cosmetic products. *100*
Figure 32 A. Meliponary owned by Mr. Nizam at the Handicraft Centre in Teluk Ewa, Langkawi; B. Mr. Nizam with his hives under the Raintree; C. & D. Ornate carved stingless beehive boxes; E. Experimental design of local roof for hive box; Tetrigona apicalis nest entrance. *101*
Figure 33 A. & B. Hives under raintree; C. & D. Hives under netted shade structure *102*
Figure 34 The main products and by-products from Madu Kelulut Resources. *103*
Figure 35 Metabolite profile analysis using Nuclear Magnetic Resonance (Bruker Ascend 700 MHz); *104*
Figure 36 The Penang Jewish Cemetery *107*
Figure 37 Entrance gate to the Jewish Cemetery *108*
Figure 38 Perimeter wall of the Jewish Cemetery *108*
Figure 39 Penang Free School *109*
Figure 40 The Penang Free School field *109*
Figure 41 The Penang Botanical Gardens *110*
Figure 42 Tetrigona apicalis (Smith, 1857) *119*
Figure 43 Lepidotrigona terminata preferred forage *123*
Figure 44 Heterotrigona bakeri (Cockerell, 1919) hairless scutellum *125*
Figure 45 Ledang expedition team *143*
Figure 46 Top left: Razif examining specimen; Top right and bottom left: entrance tubes discovered; Bottom right: Dr. Bänziger checking out the entrance tube *144*
Figure 47 Tetragonilla atripes entrance tube (Photo by Jeffry) ; members of the expedition: seated 1st row from left, Mohd Noor (Malaysia Genome Inst) Dr, Hans Bänziger (Chiang Mai University), Abu Hassan Jalil (Akademi Kelulut Malaysia), Mohd Zaki (AKM), Mohd Razif Mamat (MGI), ; 2nd Row from left, Ms. Anna (UKM), Sangdao Bänziger Halimah (MGI) , Intan & Jeffry Iskandar Foo (AKM) *144*
Figure 48 Team hard at work *145*
Figure 49 Relentless and determined *146*
Figure 50 Found something and yet to be detailed *147*
Figure 51 Collective Labels of S. moorei; T. iridipennis; T. atripes & Geniotrigona thoracica *148*
Figure 52 Stingless bees of Ranau, Sabah *152*
Figure 53 Merlimau Kelulut Honey Farm *166*
Figure 54 Qualified staff *167*
Figure 55 Harvesting Honey in full attire. *168*
Figure 56 Meliponiculture in coconut plantation *168*
Figure 57 Visit Melaka Year 2019 – VIP formal visit *169*
Figure 58 Participants of Advanced Meliponiculture course *170*
Figure 59 PIONEERING INTEGRATED BIOECONOMICS Tags *171*
Figure 60 Meliponine Repository at the Malaysia Genome Institute (MGI) *187*
Figure 61 MGI Core Facilities *188*
Figure 62 Best Entrances at MGI Repository Photo by Chui Xiao Siong *190*
Figure 63 Importance of Meliponine Repository *191*
Figure 64 MSI 15038 (2015) Metabolite Profiling Poster *191*
Figure 65 H. itama Metabolite Profile *192*
Figure 66 RapidRAW product Prototype poster *193*
Figure 67 RapidRAW Reaction Spectrum - Screening Test Kit; Previous packing above and New packing below *194*
Figure 68 SDS PAGE GEL *195*
Figure 69 Application of DNA Barcoding in Identifying Stingless Bee Colonies - MSI 16046 (2016) poster *196*

Figure 70 Phylogeny tree *197*
Figure 71 DNA Barcoding *198*
Figure 72 Fraser Hill Nature Camp Photo Gallery *207*
Figure 73 Nature & Science at Fraser's Hill *208*
Figure 74 History of beekeeping in Malaysia *234*
Figure 75 Bee products *235*
Figure 76 The log collection centre where wild colonies are sold to hot and eager customers, clearly shows a lack of food sources for the bees in transit. *243*
Figure 77 Logs at newly arrived at transit stations with absolutely no resource provided *244*
Figure 78 Recklessly stacked of newly smuggled log hives, poorly prepared for quick colony disposal to buyers. *246*
Figure 79 Preferred wrapping material *247*
Figure 80 Newly cut log hives poorly wrapped and stacked onto a trailer *247*
Figure 81 Transporting log logs in midday under the hot sun. *247*
Figure 82 Reckless cutting and shaving of log hives. *248*
Figure 83 The farm area is clean but too dense and hot. *249*
Figure 84 Farming in close quarters and narrow residential areas—livestock in low lying areas prone to flooding by the unprepared hobbyists. *249*
Figure 85 Damage done by a Malaysian Sun Bear *250*
Figure 86 Beekeepers need to avoid placement of stock in the area of frequent fogging of mosquitoes. *251*
Figure 87 This topping box is too big. *252*
Figure 88 "Creative' keepers painting boxes to preserve the hive boxes but using the wrong type of paint . *254*
Figure 89 recommended acrylic paint finish – before & after *255*
Figure 90 Hive boxes finished in water-based whitewash from late 2000 still going strong to date. *255*
Figure 91 Stingless Bee Distribution worldwide (Schwarz 1939) *256*
Figure 92 Selection of diverse nest entrance tubes - Hard resinous tubes is Dipterocarp resin sourced. *257*
Figure 93 Dipterocarp winged seeds *257*
Figure 94 Engkabang (Shorea sp.) *258*
Figure 95 Sweet Shorea (Shorea roxburghii G. Don) *258*
Figure 96 Tress favoured by Stingless bees *259*
Figure 97 Common Trees producing resin *260*
Figure 98 Stingless Bees Commonly bred in Malaysia *261*
Figure 99 Unhygienic practices *266*
Figure 100 Planting and vegetation strata levels *298*
Figure 101 Example Beescape compositions *299*
Figure 102 Poor or lack of groundcover vegetation can harbour fungus and bacteria in the decomposing litter. *301*
Figure 103 Perimeter planting to restrict bee flight range *302*
Figure 104 Foraging peculiarities *303*
Figure 105 collect pollen scattered in the leaves *304*
Figure 106 Licking up mucilage from nestmate. *304*

Appendix A - Glossary on Pollination terminology

*Bulleted entries are excerpts from http://www.fao.org/pollination/resources/glossary/en/

- **abiotic pollination**
- The movement of pollen grains from plant to plant by abiotic vectors. Wind pollination is the most common form of abiotic pollination, although water is used as a vector in rare cases by aquatic plants. (Buchmann & Nabhan 1996)

1. **Allogamy**

Allogamy is the deposition of pollen grains from the anther of the one flower on the stigma of another flower, either in the same plant or in a different plant of the same species - www.researchgate.net

- allogamy
- Fertilization between pollen and ovules of different flowers. (Richards 1997)
- **allophily**
- The condition of flowers lacking morphological adaptations to direct pollinators, such that short-tongued insects can easily use them. (Kearns and Inouye 1993)
- **allotropy**
- The condition of flower-visiting insects is only slightly adapted for pollination, lacking structural adaptations and flower constancy. (Kearns and Inouye 1993)

2. **Anemophily** or **Wind-pollination**

Anemophily or wind pollination is a form of pollination whereby pollen is distributed by wind. Almost all gymnosperms are **anemophilous**, as are many plants in the order Poales, including grasses, sedges and rushes.- en.wikipedia.org

- **anemophily**
- Wind-loving. The pollination of individual flowering plants and gyn-inosperms by the wind, an abiotic form of pollination. (Buchmann & Nabhan 1996)
- **anthophilous**
- Flower-loving; refers to animals that visit flowers. (Kearns and Inouye 1993)

3. **Anthophile**

Anthophile (plural **anthophiles**) (zoology) An organism that visits flowers. An insect is typically found on flowers or which feeds from flowers. Origin. Early 19th century; earliest use found in Encyclopaedia Britannica.

- **assortative pollination**
- Mating amongst individuals with similar characteristics (e.g. same flower colour, height, time of flowering etc.).

4. **Autogamy**,

Autogamy pollen is transferred to the stigma of the same flower - Self-pollination - Wikipediaen.wikipedia.org

- **autogamy**
- Self-fertilisation within a flower without the need for a pollinator.
- **auto pollination**

- Self-pollination within a flower resulting from the position of anthers and stigmas, or the change in position of these structures as the flower ages. (Kearns and Inouye 1993)
- **biotic**
- Regarding the actions or consequences of living species (e.g. insect, bird etc.)
- **buzz pollination**
- A pollination mechanism involving the vibration of flowers (usually with poricidal anthers) by bees as a way of obtaining pollen; the vibrations are produced by shivering of the flight muscles and are often audible to an observer. (Kearns and Inouye 1993), Thus, buzz pollination is a specialized form of pollen harvesting and pollination used by many bees (but not honeybees) to extract pollen from the pored anthers of many flowering plants, for example, blueberries, cranberries, e.g. plants, tomatoes, and deadly nightshades. About 8 % of the worlds flowering plants exhibit this form of pollination, sometimes called sonication.
- **caging**
- A technique for determining pollinating agents by surrounding plants with a screen cage prevents floral visitation by large pollinators.
- **cantharophily (cantharogamy)**
- The floral syndrome involving pollination by beetles. (Kearns and Inouye 1993)
- **chalazogamy**
- The entry of the pollen tube through the chalaza of the ovule. (Roubik 1995)

5. **Chasmogamy**.

Chasmogamy is a plant reproductive mechanism in which pollination occurs in chasmogamous flowers. Chasmogamous flowers are commonly showy, with open petals encircling exposed reproductive parts. Chasmogamous stems from Greek for "open marriage", named after the open arrangement of floral structures. - en.wikipedia.org

- **chasmogamy**
- Opening of the perianth exposes the stigma to pollen from external sources (as opposed to cleistogamy); thus, flowers are pollinated when open. (Kearns and Inouye 1993)

6. **Chiropterophily**

Chiropterophily is the pollination of plants by bats. Bat pollination is most common in tropical and desert areas that have many night-blooming plants. Nectar bats have long tongues and can hover like hummingbirds.- www.honeybeesuite.com

7. **Cleistogamy**.

Cleistogamy self-pollination and fertilization of an unopened flower, as in the flowers of the violet produced in summer - www.dictionary.com

- **cleistogamy**
- Having flowers that are self-fertilized without opening (close-fertilization). (Roubik 1995)
- **crop pollination**
- The process of pollinating (fertilising) crop species to ensure yields are produced.
- **cross-pollination**
- The transfer of pollen from the anthers of one plant to a recipient stigma on another plant may result in fertilization and fruit set. Also known as outcrossing or xenogamy. (Buchmann & Nabhan 1996)

8. **Cross-pollination**

Cross-pollination is the process of applying pollen from one flower to the pistils of another flower. Pollination occurs in nature with the help of insects and wind. This process can also be done by hand to produce offspring with desired traits, such as colour or pest resistance. colour or pest resistance.- swww.sciencedirect.com

Three types of Cross-Pollination

I. Zoophilous flowers– In this type of pollination, the pollinating agents are animals like human beings, bats, birds etc.
II. The agency of wind pollinates anemophilous flowers– These flowers.
III. Insects pollinate entomophilic flowers– These flowers.

www.toppr.com › sexual-reproductio-in-flowering-plants

- **cryptic self-fertility**
- The phenomenon in which pollination with loads of pure self-pollen rarely or never results in fruit production, but pollination using mixtures of self-and cross-pollen produces fruits with considerable numbers of selfed seeds. (Kearns and Inouye 1993)
- **cryptic self-incompatibility**
- A condition in which self-pollen tubes grow more slowly than outcross tubes and do not generally fertilize ovules when outcross pollen is present. (Kearns and Inouye 1993)
- **deception**
- (Of flowers). Attracting pollinators either by appearing to offer rewards that are not provided (e.g., resembling flowers that offer pollen or nectar, resembling oviposition sites, mates, etc.) (Dafni 1984).
- **dehisce**
- (Of anthers). To open to release pollen. (Kearns and Inouye 1993)
- **dioecious**
- Having either male flowers or female flowers on each plant in the same population. (Bernhardt 1999)
- **distyly**
- A floral polymorphism in which two morphs (pin and thrum) are produced differs in their reproductive parts' anatomy. Pin flowers have long styles and anthers placed low in the corolla tube; thrum flowers have short styles and anthers placed high in the corolla tube. Successful pollination in such species requires a crossing of the two morphs. (Kearns and Inouye 1993)
- **dystropy**
- (Of flower visitors). Not adapted or counter-adapted for utilising flowers; visits are usually disorganised, destructive, and not related to the flower's structure but may still cause pollination.

9. **Entomophily** or **Insect pollination**

Entomophily is a form of pollination whereby the pollen of plants, especially flowering plants, is distributed by insects. - en.wikipedia.org

- **entomophily**
- The floral syndrome involving pollination by insects. (Kearns and Inouye 1993)
- **ethodynamic pollination**

- Cases in which insects load pollen onto certain parts of their bodies and then deposit it on a stigma, e.g., in some Yucca and Ficus species (Galil 1973). Contrast with topocentric pollination. (Kearns and Inouye 1993)
- **euphily**
- (Of flowers). Adapted for highly specialized pollinators. (Kearns and Inouye 1993)
- **eutherophily**
- The floral syndrome involving pollination by placental mammals. (Armstrong 1979)
- **eutropy**
- (Uncommonly used). Flower-visiting insects that are completely adapted for pollination, with highly developed morphological adaptations and good flower constancy. (Kearns and Inouye 1993)
- **explosive pollination**
- Pollination is caused by the violent movement of anthers/stamen and style, alone or together with restraining petals, triggered by a pollinator's visit. (Kearns and Inouye 1993)
- **extrorse (of anthers)**
- Of an anther which dehisces away from the centre of the flower. (Richards 1997)
- **facultative autogamy**
- The occurence of cross-pollination in normally self-pollinating species. (Kearns and Inouye 1993)
- **facultative xenogamy**
- The breeding system of flowers can be cross-pollinated if pollinators are present but will self-pollinate in their absence. (Kearns and Inouye 1993)
- **flagelliflory (pendulifiory)**
- The production of flowers or inflorescences at the end of long, ropelike branches dangling down from a tree's crown presumably, facilitates access by pollinators. (Kearns and Inouye 1993)
- **floral calendar**
- A listing of flowering periods/times for a particular region, with information on the abundance and value to pollinators visiting the flowers, are blooming at any time. (FAO 1995)
- **floral constancy**
- Persistence of a pollinator in looking for flowers of one species. (FAO 1995)

10. **Geitonogamy** /ˌɡʌɪtəˈnɒɡəmi/

Geitonogamy is the fertilization of a flower by pollen from another flower on the same (or a genetically identical) plant - Oxford Languages

- geitonogamy
- Interflower pollination on an individual plant. (Kearns and Inouye 1993)
- gender adjustment
- A change in allocation to male or female investment in a system where there is variation along a broad continuum of sex investment ratios (Lloyd and Bawa 1984). (Kearns and Inouye 1993)
- geoflorous
- Flower heads are borne near ground level. (Kearns and Inouye 1993)
- **gynodioecious**

- A group of plants with both hermaphroditic and gynoecious individuals; female morph is unisexual and male morph is ambisexual. (Kearns and Inouye 1993)
- **gynodioecy**
- Where female and hermaphrodite genets coexist. (Roubik 1995)
- **gynoecious**
- Producing only or predominantly pistillate (female) flowers. (Roubik 1995)

11. Hand-pollination or **Mechanical pollination**

Hand-pollination, also is known as mechanical **pollination,** is a technique that can be used to **pollinate** plants when natural or open **pollination** is either undesirable or insufficient. This pollination method is done by manually transferring pollen from the stamen of one plant to the pistil of another.

- **hand-pollination**
- The human pollination of flowers, usually crop plants or those growing in research plots, with fresh or stored pollen. (Buchmann & Nabhan 1996)
- **hemiphily**
- (Of flowers). Flowers that are adapted for pollination by insects with intermediate levels of specialization. (Kearns and Inouye 1993)
- **hemitropous**
- (Uncommonly used). Flower-visiting insects that are partially adapted genotypes for pollination, with weakly developed structural adaptations and an intermediate degree of flower constancy. (Kearns and Inouye 1993)
- **herkogamy**
- The spatial separation of male and female sex organs in a flower so that the flower cannot self-pollinate. (Kearns and Inouye 1993)
- **hermaphroditic**
- 1. A flower with both stamens and pistils; 2. A plant with only perfect flowers, functionally both staminate and pistiilate; 3. A group of plants containing only hermaphroditic plants. (Kearns and Inouye 1993)
- **heteranthery**
- Possession of a dimorphic androecium with two or more different types of anthers. (Kearns and Inouye 1993)
- **heterodichogamy**
- A breeding system in which two mating types are different in their temporal development pattern, such as both protandrous and protogynous individuals. (Kearns and Inouye 1993)
- **heterogamous**
- A plant has two or more kinds of flowers. (Roubik 1995)
- **heterozygosity**
- A measure of the genetic diversity (at the gene level) in a natural population. It considers the number of different gene loci (sites) on the chromosomes for different individuals. (Buchmann & Nabhan 1996)
- **homogamous**

- No difference in the timing of anther dehiscence and stigma receptivity. (Kearns and Inouye 1993)
- **homostyle**
- A flower with anthers and stigmas at the same level, as opposed to a heterostylous condition. (Kearns and Inouye 1993)
-

12. **Hydrophily**

Hydrophily is a fairly uncommon form of pollination whereby pollen is distributed by water flow, particularly in rivers and streams. **Hydrophilous** species fall into two categories: (i) Those that distribute their pollen to the surface of the water.- en.wikipedia.org ›

- **hydrophilous**
- Pollination by fresh or saltwater, as in the case of eelgrass. Since pollen is usually damaged by immersion in water, this form of pollination is rare. (Buchmann & Nabhan 1996)
- **hypanthium**
- A floral tube or cup, formed by the fused bases of the sepals, petals, and stamens. (Kearns and Inouye 1993)
- **hypochil**
- One of the three parts of the fleshy lip of some orchid flowers. (Kearns and Inouye 1993)
- **hypogynous**
- A flower in which the sepals, petals, and stamens are attached at a level below the ovary. (Kearns and Inouye 1993)

- **illegitimate pollination**
- Pollination with the wrong pollen in a heterostylous pollination system, e.g., pollen from a long stamen deposited on a short style instead of a long one. (Kearns and Inouye 1993) 2. Pollination of a flower of the same self-incompatibility genotype as the pollen parent (see legitimate). (Richards 1997)
- **inbreeding**
- Sexual reproduction involves the interbreeding of closely related individual plants through self-pollination and often backcrosses. (Buchmann & Nabhan 1996)
- **inbreeding depression**
- The loss of vigour and general health suffered by an organism resulting from inbreeding (reproduction between two closely related individuals).
- **indeterminate**
- Continuing to grow after flowering starts. (Roubik 1995)
- **indeterminate bloomer**
- A plant in which flower production is irregular and inconstant, especially in the absence of pollination (most insect-pollinated annual crops, except cereals). (FAO 1995)
- **indeterminate inflorescence**
- Inflorescence with outer or lower flowers opening first. (Kearns and Inouye 1993)
- **insectary plants**
- Plants were grown to attract beneficial insects (pollinators and predators of crop enemies) (Singh 2004)
- **intine**

- The inner of the two walls of a mature pollen grain. (Kearns and Inouye 1993)
- **introrse**
- Of an anther which dehisces towards the centre of a flower (see extrorse). (Richards 1997)
- **involucre**
- The bracts around a flower cluster or head of a composite. (Kearns and Inouye 1993)
- **isantherous**
- Of flowers. With monomorphic anthers. (Kearns and Inouye 1993)
- **isostemonic**
- With many stamens equal to the number of sepals or petals. (Kearns and Inouye 1993)
- **keel**
- A central ridge; in leguminous flowers, the two fused petals that enclose the stamens. (Kearns and Inouye 1993)
- **labellum (plural labella)**
- 1. Lip of a flower; on an orchid, what appears to be the lowermost petal (actually the upper petal morphologically, but the pedicel twists). 2. A fleshy pad at the end of a fly proboscis. (Kearns and Inouye 1993)
- **legitimate**
- Pollination of a flower of a different self-incompatibility genotype to the pollen parent, likely to result in a fertile seed-set (see illegitimate). (Richards 1997)
- **locule**
- Compartment or cell of ovary, anther, or fruit. (Roubik 1995)
- **malacophilous**
- Pollinated by invertebrate transfer.
- **male-sterile**
- Crop cultivar whose male plants/flowers do not produce viable pollen and therefore cannot fertilise.
- **mass flowering crop**
- A crop in which all plants simultaneously bloom.
- **megaspore**
- A haploid female meiospore that produces the female gametophyte by mitosis. (Kearns and Inouye 1993); Gives rise to a female gametophyte. (Richards 1997)
- **megasporophyll**
- Spore bearing leaf which bears megaspores; the ovary or carpel in flowering plants. (Richards 1997)
- **meliponiculture**
- The ritualized keeping of stingless bees (meliponines), especially in a traditional context by Mayan Indians in parts of Mexico and Central America. These bees are usually kept in hollow logs and periodically raided to obtain honey and wax. (Buchmann & Nabhan 1996)
- **meliponine bee**
- Stingless bees belong to the genera Melipona and Trigona, which occur in both the New World and Old-World tropics. They are highly social and live in populous perennial colonies. (Buchmann & Nabhan 1996)
- **melissopalynology**

- The pollen analysis of honey; commonly used to determine the geographical and floral origin of honey. (Kearns and Inouye 1993)
- **melittophily**
- The floral syndrome involving pollination by bees. (Kearns and Inouye 1993)
- **mentor pollen**
- Pollen that is mixed with an incompatible pollen type to stimulate fruit and seed development and production of some seeds fertilized by the incompatible pollen (see rescue pollination). Mentor pollen is commonly treated to destroy its fertilizing abilities or destroy the incompatibility mechanism (Brown and Adiwilaga 1991).
- **mesochil**
- One of the three pans of the fleshy lip of some orchid flowers. (Kearns and Inouye 1993)
- **mess-and-soil pollination**
- A type of pollination usually occurring in primitive flowering plants and is affected by beetles so named because the beetles' blunder around the flowers, chewing on floral parts, eating pollen, and defecating. (Buchmann & Nabhan 1996)
- **metatherophily**
- The floral syndrome involving pollination by marsupials. (Armstrong 1979)
- **microsporangia**
- Structures in which microspores are produced; in angiosperms, the elongated paired pollen sacs constitute one of the two lobes of a typical anther; thecae. (Kearns and Inouye 1993)
- **microspore**
- A haploid male spore; uninucleate structure released from tetrads after meiosis that produces the male gametophyte (pollen grain) by mitosis. (Kearns and Inouye 1993); Gives rise to a male gametophyte (includes pollen). (Richards 1997)
- **microsporocytes**
- Pollen mother cells contained within the microsporangia (thecae) that undergo meiosis to produce haploid microspores. (Kearns and Inouye 1993)
- **microsporogenesis**
- The production of microspores by meiosis. (Kearns and Inouye 1993)
- **microsporophyll**
- A leaflike structure that bears microsporangia. (Kearns and Inouye 1993)
- **mixed inflorescence**
- A branched inflorescence with both racemose and cymose components, as in grape and mango, in which the main inflorescence axis is racemose, the small ultimate branches cymose. (Roubik 1995)
- **monocarpic**
- Flowering only once during a lifetime; semelparous. (Kearns and Inouye 1993)
- **monoclinous**
- Bearing hermaphrodite flowers. (Kearns and Inouye 1993)
- **monoecious**
- Having two different kinds of imperfect flowers (male and female) on each plant in the same population. (Bernhardt 1999) as in the squashes, gourds, and pumpkins, for example.
- **monolectic**

- Bees collect pollen from only one species of plant to provision their brood cells. The bee will usually collect nectar from a wider range of plants. (Shepherd, Buchmann et al. 2003)
- **monolecty**
- The pollen-collecting behaviour of certain solitary bee species that, throughout their range and year to year, specialize on the pollen of just one species or related congeners. See oligolecty and polylecty. (Buchmann & Nabhan 1996)
- **monophilic**
- (Of flowers). Pollinated by one or a few related species of pollinators. (Kearns and Inouye 1993)
- **monotelic inflorescence**
- One in which the apex of the inflorescence axis ends with a terminal flower. (Kearns and Inouye 1993)
- **monotropic**
- Insects that visit a single plant species. (Kearns and Inouye 1993)
- **myophily**
- The floral syndrome involving pollination by flies. (Kearns and Inouye 1993)
- **myrmecophily**
- The floral syndrome involving pollination by ants. (Kearns and Inouye 1993)
- **nectar**
- A secretion produced by flowers to attract pollinators usually containing sugars, amino acids, and other compounds that are of nutritional importance to flower visitors. (Kearns and Inouye 1993)
- **nectar breeding**
- Selective breeding of crops for plants with high nectar production. (FAO 1995)
- **nectar corridor**
- Habitat along the migration route of a pollinator, either a continuous strip or stepping stones, which provides foraging and refuelling resources across an otherwise inhospitable landscape. (Shepherd, Buchmann et al. 2003)
- **nectar guide**
- A marking in a contrasting colour is not always visible to humans, which serves as an orientation cue to a flower visitor of where to find nectar. (Kearns and Inouye 1993)
- **nectar reward**
- The sugar secretions are produced by flowers as a reward for visiting pollinators.
- **nectar robbing**
- Biting into a flower to extract the nectar illegitimately and usually without affecting pollination. A behaviour exhibited most commonly by some bees, but also by ants, hummingbirds, and other animals. (Kearns and Inouye 1993)
- **nectar theft**
- Obtaining nectar in a fashion other than that for which a flower has evolved and without affecting pollination, but not by causing mechanical damage to the flower (e.g., bypassing the reproductive parts of a flower with a long, thin proboscis). (Kearns and Inouye 1993)
- **nectariferous**
- Producing nectar. (Kearns and Inouye 1993)
- **nectary**

- A specialized region of floral tissue, usually at the base of the innermost floral tube, where nectar, a mixture of sugars and amino acids in water, is secreted. The nectar solution forms pools in this region, and pollinators drink from it. (Buchmann & Nabhan 1996)
- **nototribic**
- Zygomorphic flowers with stamens and style are placed so that they come into contact with the forager's body's dorsal surface. (Kearns and Inouye 1993)
- **obligate outcrosser**
- For reasons of physiological incompatibility, a flowering plant species must receive and donate pollen in the form of cross-pollination. It cannot pollinate itself between flowers on the same plant. Greater seed set levels are achieved with outcrossing and increased genetic recombination (mixing of genes). (Buchmann & Nabhan 1996)
- **oligandry**
- Possessing few stamens. (Kearns and Inouye 1993)
- **oligolectic (of flower visitors)**
- Bees that collect pollen from a limited range of plants (often those in a single genus or family) to provision their brood cells. These bees will usually collect nectar from a wider range of plants. (Shepherd, Buchmann et al. 2003)
- **oligolege**
- A flower visitor that specializes in a few related plant species. (Kearns and Inouye 1993)
- **oligophilic**
- (Of flowers). Pollinated by a few related animal taxa. (Kearns and Inouye 1993)
- **oligotropic**
- (Of flower visitors) Visiting only a few related host plants. (Kearns and Inouye 1993)
- **ombrogamy**
- Pollination by rain. (Kearns and Inouye 1993)
- **open-pollinated flower**
- Pollination by native pollinators; normal, natural pollination without additional pollen input from humans (Winfree et al. 2007)

13. **Ornithophily**

Ornithophily is pollination by birds (such as hummingbirds, sun-birds, honey-eaters, etc.) - doubtnut.com › question-answer-biology

- **ornithophily**
- Pollination of bird-adapted flowers by perching and hovering birds. Usually, these flowers are large and colourful and sturdily built, so the birds probing bills and feet do not damage the ovaries within the ovaries. (Buchmann & Nabhan 1996)
- **osmophore**
- Scent-producing gland on a flower. (Kearns and Inouye 1993); Floral organ, often formed from a petal, which is long, narrow, drooping, often dark, and usually foul-smelling, usually as part of a myophilous syndrome. (Richards 1997)
- **outcrossing**
- The chief means of reproduction by flowering plants. Such plants can only be fertilized by pollen from other plants. (Buchmann & Nabhan 1996)
- **phalaeonophily**

- The floral syndrome involving pollination by small moths. (Kearns and Inouye 1993)
- **pioneer pollen**
- Pollen is applied first in rescue pollination techniques. This pollen carries the genotype desired in the cross but is normally not cross-compatible with the ovules. (Kearns and Inouye 1993)
- **pistil**
- The female reproductive organ of a flower is further subdivided into an ovary, style, and stigma. (Buchmann & Nabhan 1996). The normally central, seed-producing part of a flower. (Roubik 1995)
- **pistillate**
- Functionally female flowers, with pistils only. (Kearns and Inouye 1993)
- **pistillate sorting**
- Processes occurring in the pistil that result in nonrandom seed parentage to the pollen deposited on the stigma (Bertin et al. 1989)

14. **Pollination**

Pollination is the act of transferring pollen grains from the male anther of a flower to the female stigma. The goal of every living organism, including plants, is to create offspring for the next generation. One of the ways that plants can produce offspring is by making seeds. - USDA Forest Service www.fs.fed.us › wildflowers

15. Pollinator or **Pollinating agent**

A **pollinator** is an agent that causes that transfer. Pollinators range from physical agents, especially the **wind (wind** pollination is called anemophily), or biotic agents such as **insects**, **birds**, bats and other animals (pollination by **insects** is called entomophily, by **birds** ornithophily, by bats chiropterophily).

- **pollen**
- The powdery grains produced by angiosperm anthers or the microsporangia of gymnosperms, which contain the nucleus that fertilizes the oosphere to form a seed. (Roubik 1995). Each pollen grain contains a tube cell connected to one or two sperm cells. These cells are encircled by a tough, double-layered wall. (Bernhardt 1999).
- **pollen brush**
- A patch of special hairs on a bee's body that is used to carry dry pollen back to the nest. Usually, the scopa is on the back legs, although it is on the sides of the thorax or the underside of the abdomen in some families. (Shepherd, Buchmann et al. 2003)
- **pollen carryover**
- pollen carryover. The deposition of pollen beyond the first flower after the one where it originated. (Kearns and Inouye 1993)
- **pollen coat**
- A layer of fat globules and protein film that clings to the surface of most pollen grains. Often contains pigments and scent molecules that give the grains distinctive colour and scents. (Bernhardt 1999)
- **pollen flow**
- The movement of pollen grains to flowers other than those where they originated. (Kearns and Inouye 1993)

- **pollen flower**
- Flowers that offer visitors abundant pollen but little or no nectar. (Proctor et al. 1996)
- **pollen grain**
- The multicellular male gametophyte. (Kearns and Inouye 1993)
- **pollen limitation**
- Production of less than the maximum seed potential due to inadequate numbers of pollen grains to fertilize ovules. (Kearns and Inouye 1993). The degree to which pollinator services are sufficient to maximise fruit/seed set. (Burd 1994). An inappropriate or insufficient deposition of pollen on the stigma (sensu Ashman et al. 2004, Knight et al. 2005).
- **pollen mother cells**
- The cells that divide produce microspores that mature into pollen grains. (Kearns and Inouye 1993)
- **pollen presenter**
- A specialized region of the style in the genus Banksia (Proteaceae) onto which pollen dehisces from the anthers just before the flower opens. (Kearns and Inouye 1993)
- **pollen resource**
- The pollen inside a flower as a resource for flower visitors.
- **pollen robbing**
- Biting into a flower to remove pollen, usually without affecting pollination. The behaviour exhibited by some species of bees and perhaps other animals. (Kearns and Inouye 1993)
- **pollen specialization**
- (Of flower visitors). A preference or adaptation for utilising pollen from only a certain species or type of flower.
- **pollen spraying**
- Mechanical pollination technique of spraying plants with a pollen solution through an airbrush apparatus. (Calzoni & Speranza 1998).
- **pollen theft**
- Obtaining pollen in a fashion other than that for which a flower has evolved and usually without effecting pollination, but not by causing mechanical damage to the flower (e.g., gathering of pollen from a large flower by small bees that will not affect pollination). (Kearns and Inouye 1993)
- **pollen trap**
- Installation on bee-hive entrances to collect pollen from bees entering the hive. (FAO 1995)
- **pollen tube**
- A thin tubular outgrowth of pollen grain usually contacts the stigma, which penetrates style to the ovary, permitting sperm nuclei to unite with an egg cell. (Roubik 1995)
- **pollen viability**
- A pollen grain's ability to successfully fertilise (FAO 1995); sterile or aborted, or unviable pollen will not result in pollination while viable pollen will.
- **pollen/ovule ratio**
- The number of pollen grains produced by a flower divided by the number of ovules; can be used as an indication of the breeding system.
- **pollenizer**
- Plant source of pollen for fertilizing receptive stigmas. (Roubik 1995)

- **pollenkitt**
- The adhesive surface of pollen grains; can contain lipoidal or pigmented substances that may add colour or odour to the pollen and may cause pollen grains to adhere to each other during dehiscence or to pollinators. (Kearns and Inouye 1993)
- **pollinarium**
- In orchids and asclepiads, the structure detached from the male plant bearing sacs of the pollen grain united to an adhesive disc that attaches to the pollinator. (Roubik 1995)
- **pollinating**
- Transferring pollen from anthem to stigmas. (Roubik 1995)
- **pollination**
- The process of moving pollen from one flower's anthers to the stigma of another or the same flower. Equally vital processes of fertilization and seed set follow from pollination. Pollination can be affected either by abiotic means such as gravity, wind and water, or animals such as bats and bees. (Buchmann & Nabhan 1996)
- **pollination agent**
- Particular vector for moving pollen from male to female plant reproductive parts; the agent of pollination.
- **pollination biology**
- The study of all aspects of pollination, including plant/pollinator interactions, pollen types, pollination mechanisms, flower morphology, breeding systems etc.;
- **pollination brush**
- An apparatus used to pollinate pears in China: a brush made from cigarette filters or chicken feathers attached to a stick for ease of application (Ya et al. n.d.)
- **pollination decline**
- This refers to a decline in the rate that plants are pollinated due to increasing habitat fragmentation and modification. (Cunningham 2000)
- **pollination deficit**
- A quantitative or qualitative deficit of pollination that results in a decrease in the sexual reproductive output of plants. (Wilcock & Neiland 2002)
- **pollination ecology**
- The study of the ecological and evolutionary relationships involved in the pollination process. (Buchmann & Nabhan 1996)
- **pollination intensity**
- The number of viable pollen grains that are deposited on a stigma. (Kearns and Inouye 1993)
- **pollination interaction**
- The reciprocal action, effort or influence between a plant and the organism provides a pollination service.
- **pollination limitation**
- Absence of pollen deposition (Wilcock & Neiland 2002); a difference between the naturally occurring pollination level - that is, open pollination and the pollination that resulted from various treatments that could be interpreted as improved pollination. (Vaissere et al. 2008)
- **pollination management plan**

- An agreed set of practices to be undertaken within a defined landscape area by a community of farmers or land managers to promote pollinators and thus increase the level of pollination services to target crops. (FAO 2009)
- **pollination mode**
- Method of pollination, either selfing or outcrossing. (FAO1995)
- **pollination services**
- Pollination acts are performed by all the various animals that dependably visit certain species of flowering plants. (Buchmann & Nabhan 1996)
- **pollination stick**
- An apparatus used to pollinate pears in China: a stick to which a pollination brush is attached (Ya et al. n.d.)
- **pollination syndrome**
- A set of plant attributes indicative of pollination via a particular vector (i.e. wind, bee, bird etc.) (Mayfield 2008). An old concept, dating back to the early works of Europeans including Paul Knuth and Hermann Mueller, suggesting that the floral visitors a flower receives (hummingbird, bat, fly, butterfly, moth, bee, beetle) can be predicted from the flowers suite of morphological characteristics and rewards. Although subject to misapplication, the concept is still of great value in teaching about pollination. (Buchmann & Nabhan 1996)
- **pollinator**
- An animal that moves compatible pollen to a receptive stigma of the same plant species, such that fertilization and seed production can occur. (Roubik 1995)
- **pollinator abundance**
- The number of individual pollinators in a particular ecosystem, community or visiting a particular plant population (Mayfield 2008). A population characteristic affects the number of visits a plant receives from a given pollinator species (Young 1988).
- **pollinator diversity**
- A general term that can refer to the number of pollinator species found in a particular area or any other diversity measure or diversity index, which may include information on the abundance of each species as well as the number of species. This is most commonly a spatially explicit measure.
- **pollinator effectiveness**
- A measure of the accomplishments of a single visit of an individual animal in terms of pollen deposited, pollen removed, seeds produced, percentage of available florets pollinated, etc. (Spears 1983, Young 1988, Neff and Simpson 1990). (Kearns and Inouye 1993)
- **pollinator efficiency**
- A measure of both the costs (flower damage, pollenate, etc.) and benefits (pollen deposited, pollen removed, seeds produced, etc.) of a single visit of an individual animal (Neff and Simpson 1990). (Kearns and Inouye 1993)
- **pollinator garden**
- A garden specifically designed to attract butterflies, hummingbirds, or bees. (Buchmann & Nabhan 1996)
- **pollinator importance**
- A measure combining the pollinator abundance and pollinator effectiveness (Young 1988). (Kearns and Inouye 1993)

- **pollinator limitation**
- When pollinators are a limiting factor for the maximization of reproductive success in plants, pollinators are limiting when they occur in insufficient numbers to produce maximal fruit set in an insect-pollinated plant.
- **pollinator mortality**
- The increase in the rate or abundance of pollinator deaths in a particular ecosystem. Frequent causes are excessive or misused pesticide application or disease.
- **pollinator movement**
- Refers to the pattern and direction of movements of a particular pollinator across the flowers of a particular plant community.
- **pollinator network**
- A network between communities of pollinators and plants (2-mode network) or between species of pollinators or between plants in a community (1-mode network). (Olesen et al. 2006)
- **pollinator pool**
- All of the pollinators that visit a certain plant in a given area. (Buchmann & Nabhan 1996)
- **pollinator syndrome**
- A system used by many floral biologists during the middle of the twentieth century to describe certain groups of flowers classified according to their physical attributes (shape, size, open or tubular, and soon) as well as their colour, odours, and types and amounts of floral rewards offered. Thus deep-throated red flowers with no scent and abundant nectar are most often pollinated by hummingbirds in the New World tropics. Therefore such blossoms are said to be hummingbird-pollinated and belong to this pollinator syndrome. Similarly, one can describe floral guilds that are adapted by pollination by flies, beetles, birds, butterflies, moths, and bats. (Buchmann & Nabhan 1996)
- **pollinator-limited**
- A system in which pollinators- and therefore pollen grains- are in short supply and limit the number of fruits and seeds that a local plant population can produce. (Buchmann & Nabhan 1996)
- **pollinium (pl. pollinia).**
- A mass of pollen grains united into a single dispersal unit for more efficient transport by a specialized pollinator. Pollinia are routinely found in the milkweed and orchid families. This pollen is usually quite safe from ingestion by pollinators. (Buchmann & Nabhan 1996)
- **pollinizer**
- A male plant used solely for pollen production within a fruiting crop, e.g. kiwifruit; dates.
- **polygamous**
- Bearing both unisexual and bisexual flowers at the same time. (Bernhardt 1999)
- **polygamy**
- Having both perfect flowers and those of one sex (staminate or pistillate). (Roubik 1995)
- **polylecty**
- (Of flower visitors). The pollen-collecting and feeding habits of social bees and certain solitary bees that routinely gather pollen from unrelated plant families and genera in one area and over a long time. One of the most polylectic species is the introduced honeybee, Apis mellifera, for it has the broadest pollen diet. See monolecty and oligolecty. (Buchmann & Nabhan 1996)
- **polylege**

- A generalist flower visitor. (Kearns and Inouye 1993)

- **rescue pollination**
- Application of a second pollen load of a different genotype to promote seed formation from the first pollen applied. E.g., pollination of cultivated tetraploid potato flowers with pollen from a triploid hybrid does not result in a seed set. However, if this first pollination is followed by pollination with pollen from cultivated tetraploids, fruit development is normal, and some seeds are derived from the first pollination (Brown and Adiwilaga 1991).
- **robbing**
- In pollination ecology, the destruction of at least part of a flower to obtain nectar, pollen or another resource. (Roubik 1995)
- **rugose**
- Roughly sculptured, as of the exine of a pollen grain. (Richards 1997)
- **sapromyiophily**
- The floral syndrome involving pollination by carrion- and dung- flies. (Kearns and Inouye 1993)
- **secondary pollen carryover**
- Transport of previously deposited pollen grains from stigma to another flower by a vector. (Kearns and Inouye 1993)
- **secondary pollen presentation**
- Deposition of pollen, typically before anthesis, onto some floral part such as the style, from which it is then transferred to flower visitors (E.g., some Rubiaceae and Proteaceae). (Kearns and Inouye 1993)

16. **Self-pollination** /sɛlfpɒlɪˈneɪʃ(ə)n/

Self-pollination is the pollination of a flower by pollen from the same flower or another flower on the same plant. - Oxford Languages

- **Self-pollination** can lead to inbreeding depression caused by the expression of deleterious recessive mutations or the reduced health of the species due to the breeding of related specimens. An anti-selfing mechanism is self-incompatibility, a plant's ability to reject its pollen and that of closely related individuals.
- **self-compatible**
- A plant is capable of being fertilized by pollen within its group of flowers (inflorescence). (Roubik 1995)
- **self-fertile**
- A flower or floret is capable of being fertilized by its pollen. (Roubik 1995)
- **self-incompatible**
- Flowering plants that are incapable of pollinating themselves rely on genetically distinct pollen from distant plants or nonrelatives to form fruits and seeds. (Buchmann & Nabhan 1996)
- **self-pollinating**
- Autogamous; capable of placing its pollen upon its stigma. (Roubik 1995)
- **self-sterile**
- Incapable of becoming fertilized by its pollen. (Roubik 1995)

- **selfing**
- Fertilization of an ovule by a pollen grain of the same genet. (Roubik 1995)
- **semi-compatibility**
- Where two genets share some but not all gametophytic compatibility traits thus, in some crosses, some pollen grains can affect fertilization, and others cannot. (Roubik 1995)
- **sphingophily**
- The floral syndrome involving pollination by hawkmoths. (Kearns and Inouye 1993)
- **sporophytic**
- Of the sporophyte (usually diploid) generation, in the seed plants, the dominant generation, the genet that develops from a zygote, is the sporophyte; in sporophytic incompatibility, it is the genotype of the anther that produces the significant pollen grain, the anther being of the sporophyte generation. (Richards 1997)
- **sporophytic incompatibility**
- An incompatibility system in which the pollen donor plant's genotype determines the expression of pollen grain compatibility type. (Kearns and Inouye 1993)
- **sporopollenin**
- The resistant material constituting the exine; believed to be derived from the polymerization of carotenoids. (Kearns and Inouye 1993)
- **syngamy**
- Union of gametes to form a zygote. (Kearns and Inouye 1993)
- **therophilous**
- The floral syndrome involving pollination by mammals. (Kearns and Inouye 1993)
- **thievery**
- In pollination ecology, removal of a floral reward by an animal, where pollination does not follow (as opposed to robbing). (Roubik 1995)
- **thysanopterophily**
- Pollination by thrips (a word coined by David Inouye as there is no term to describe this type of pollination). (Kearns and Inouye 1993)
-
- **topocentric pollination**
- Cases in which contact between a foraging insects body and pollen and stigma is ensured by the topography of the flower; contrast with ethodynamic pollination. (Kearns and Inouye 1993)
- **trap flower**
- Flowers (e.g., Aristolochia) affect pollination by temporarily trapping pollinators in flowers. (Kearns and Inouye 1993)
- **unisexual flower**
- Flower-bearing reproductive organs of one gender. (Kearns and Inouye 1993). Flowers in which either stamens or pistils are functional, the nonfunctional member may be completely lacking, as in pistillate flowers of pawpaw, or rudimentary or aborted, as in fig, banana, some grapes, and in the staminate flowers of pawpaw. (Roubik 1995).
- **vibratile pollination**
- Buzz pollination. (Kearns and Inouye 1993)
- **virgin flower**

- Flowers that are known to have never been visited by an effective pollinator or pollen vector. (FAO 1995)
- **wild pollinator**
- A pollinator that lives and forages without the assistance or manipulation of nesting sites by humans.
- **wildflower mixture**
- The mixture of native flowering species sown in the grassy margins of arable fields has proven to increase pollinator diversity and efficiency. (Carvell et al. 2004)
- **wind-pollinated**
- Plants that depend on the wind to carry their pollen grains (Buchmann & Nabhan 1996)

17. **Xenogamy** (Greek xenos=stranger, gamos=marriage)

Xenogamy is the transfer of pollen grains from the anther to the stigma of a different plant. This is the only type of pollination that brings genetically different types of pollen grains to the stigma during pollination. - Wikipediaen.wikipedia.org

- **xeriogamy**
- Fertilization between pollen and ovules between different genets. (Roubik 1995)
- **xeromorphic**
- With phenotypic attributes enabling a plant to survive dry conditions. (Richards 1997)

18. Zoophily

Zoophily is a form of pollination whereby pollen is transferred by animals, usually by invertebrates but in some cases vertebrates, particularly birds and bats, and other animals. – en.wikipedia.org

- **zoophily.**
- Pollinated by animals. (Kearns and Inouye 1993)
- **zygomorphic**
- (Of a flower). Bilaterally symmetrical. (Kearns and Inouye 1993)
- **zygote**
- The result of the fusion of a sperm sex cell with an unfertilized egg. Following fertilization, this zygote begins to undergo cellular division and function grows into a recognizable embryo in a matter of hours or days. (Buchmann & Nabhan 1996)

Appendix B – Genera Identification Key Update

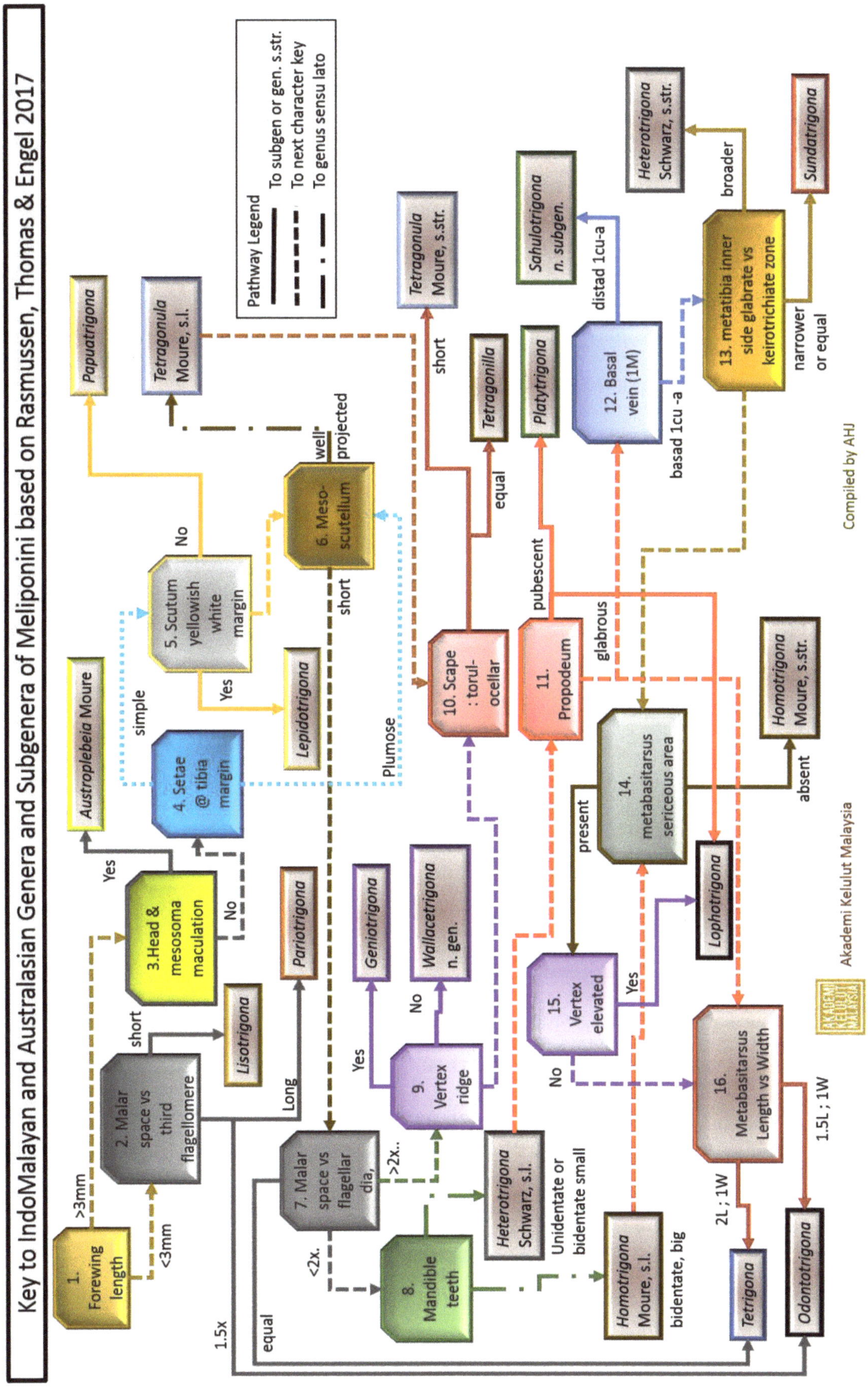

Appendix C – Types of swarms

by Abu Hassan Jalil and Nick Powell (Photo inset)

A swarm is defined as a large dense group of flying insects.

ST.1 Hovering swarm: Expand defence perimeter and vigilant to intruders. This is characteristic of a hive with no real intrusion as yet or the presence of a hatchling gyne (virgin queen). It can be from an overpopulated colony venting tidal gas exchange or detecting dummy intruders to real intruders. These hovering swarms may develop into other swarm characteristics as the dangers manifest themselves or the intrusions materialize.

ST.2 Alert Swarm: This can be when a gyne is due to hatch sensed from colony pheromones and hive activity. The alert swarm is related to the mating season as opposed to the Alarm swarm that is generally in battling mode.

ST.3 Alarm Swarm: Happens when the colony is under threat from external impending intruders or danger. These can be sparse initially but grows bigger into attack mode as the imminent danger becomes real.

Can be seen a tight ball swarm 1-4 meters square with bees' pairs falling and flying back up to join the swarm. This happens when bees grab a friendly bee and are mistaken. This is a key difference between this type of swarm and the attacking swarm. Defensive or alarm swarms can last 1-2weeks before they feel secure. This swarm is gentler and its pattern while being vortex is generally not as fast-moving as the attacking swarm

ST.4 Attack swarm: A follow-up from the Alarm Swarm where the bees come out in full force to attack the current intrusion. This defensive strategy gains its strength in numbers and turns into an offensive strategy. The attack becomes relentless when the brood or its nest produce are ravaged, e.g. bears, badgers, squirrels or primates rummaging for the honey. Other bees or insects are robbing the propolis.

Swarm contains bees moving in an erratic figure of 8 flight vortex pattern with bees falling out of the swarm in pairs. Bees carry sticky substances on legs and cling locked to enemy bees locking on with their mandibles to each other. Each pair of bees die as a result.

ST.5 Invading swarm: Take over manoeuvre by a more aggressive species over a weaker colony. Generally, occurs when a new colony is brought into a farm or due to lack or insufficient foraging sources. Sometimes even when the foraging sources are available but not particularly favoured by the newly imported species, but the indigenous species may invade to undermine the competition.

This usually starts a few days prior with the arrival of 10-20 scout bees which can be observed hovering in front of the hive entrance. These scout bees test the strength of the colony before the large-scale invading swarm appears in great numbers. The swarm can be noted landing on the hive Turning the face of the hive black with bees. Invaders that land on the hive group together and fan the distressed pheromones into the air to guild the rest of the bees to the location of the invasion. It is a numbers game, and the invaders rarely make a gamble on taking a hive.

ST.6 Mating swarm: Generally, it occurs during dry mating season. These generally constitute male bees. Different species in different regions of varied climates or seasons may have different peculiarities. Males can also bee is seen congregating at the sides of boxes or nearby pillars or tree trunks/

Drones can be observed in numbers ranging from 20 to-1000. Hovering and landing on hive roofs and branches nearby to the hive location that contains the virgin queen. These can be observed right throughout the year and are more prevalent when ants send out their nymph queens. Drone bees hold their antenna in a curved fashion, which can be likened to bunny ears. Workers will push drones away from the colony entrance and are on high alert during this time. Hive activity is heightened, and work is done in double time, with bees moving very fast in the days before the queen is mated.

ST.7 Warming swarm occurs after a spate of cold weather or a period of no sunlight exposure. The swarm usually hover over pockets of sunlight that may be near the hive. These swarms can be as large as 3 meters in width. This was found to occur in Australia after the winter season.

ST.8 Colonizing swarm: This happens when a colony has reached its capacity. Bees leave the colony in great numbers and cluster on trees, the ground and vertical surfaces in the area they wish to live. Once the swarm workers decide a location, a pheromone is splattered on the area and continually worked over until the bees in the area are in flight form together.

All bees will face one direction once the location and the swarm can be seen clustering at the point of entry. This is the only swarm that the bees will face in one direction in flight instead of a vortex or figure of 8 flight patterns. Once the location is decided, some of the bees will return to the mother hive to take some wax and propolis on their legs. This aids in the fast setup of a new home.

ST.9 Drifting Swarm: this occurs in a farm when boxes are placed too close and multiple swarms occur. The returning bees may "drift" into the wrong hives in the chaos. This was observed in T. laeviceps and *T fuscobalteta* colonies in the west end (Ujong Kulon) of Java Meliponary.

ST.10 Exploratory swarm: Occasionally seen covering large areas up to 10 meters in a span with bees searching vertically and horizontally, usually upon a wall or hedge. Bees are landing on and then to rest. Attracted to darker spots of colour variation. Often scent-marking these areas. This can be the initial exploratory move before the colonizing swarm.

ST.11 Intermingled swarm: This happens when two swarms of different species swarm in proximity simultaneously.

This swarm differs from a fighting or attacking swarm in that the bees are not in an aggressive flight pattern or locking together. The swarm has a soft flight pattern and shows much interest in scent marking. The swarm carries workers and queens of both species, which can change from day to day, leaving one more predominant in numbers than the other. Queens can be seen as larger and are a few days old before they join this swarm. Because of the multiple pheromones present in this type of swarm, they are unlikely to move into a trap hive and stay in one area for weeks. Many clusters of bees containing 10-50 bees can be seen working resinous substances along tree branches and vertical walls, posts etc. The size of this swarm can be up to 6 meters in span.

ST.12 Fighting swarms: Fighting usually occur as preludes to colonizing or developed from an attacking situation on forage competition. This predicament is when sources are limited or depleted due to extreme weather conditions in bee farms or Meliponary. Other instances are when a new colony is introduced into an already overcrowded farm.

Pairs of flying bees will grasp one another with their jaws and fall to the ground, wrestling to death. In a major fighting swarm, the result can be a carpet of thousands of dead and dying bees on the ground, locked in mortal combat (A. Dollin 2008). This is when all bees are workers, and fighting swarms can result from attacking bees that have come from an unknown hive quite some distance away. Stingless bees may produce alarm pheromones during a fighting swarm. These special scents, produced from glands in the bees' heads, may attract other bees from neighbouring hives to join in the fight. Ros Gloag's experiments on *Tetragonula carbonaria* in 2008 showed they could have congregated from 7 different hives.

Although each colony has their nestmate recognition pheromones, they sometimes, in the heat of battle, lock jaws with nestmates but release them as soon as they fall to the ground.

Index

abaxially, 21
Abraham, 131, 135
Acai berry, 20
Actinopyga sp, 100, 105
Aedes, 242
aesthetically pleasing, 296
Africa, 41
Ahoerstemeier, 249
air-conditioned vans, 237
Akademi Kelulut Malaysia, 12, 82, 31, 133, 169, 303
AKM, 31, 169, 170, 303, 305
Allen-Wardell, 13
allogamous, 75
allspice, 17
aloe vera, 298
Alor Star, 133
American chestnut, 43
amino acids, 319, 320
AMNH, 114, 305
Ampang, 132, 137, 139
Anacardium occidentale, 37, 42
Ananas comosum, 12
Ananas comosus, 23, 37, 63
Angsana trees, 112
angular shape, 144
Annona, 37, 68
Annona muricata, 37, 68
Anthesis, 12, 21
Antigonon leptosus, 149
Antioxidant, 107, 108
ants, 59, 66, 83, 141, 319, 333
aphids, 298
Apiary, 37
apicalis, 122, 137, 144, 142
Apiculture, 82
Apidae, 301, 302, 304, 305
Apis bees, 23, 82
Apis dorsata, 66
Apis mellifera, 71, 75, 305, 327
arabica coffee, 149
Arbitration, 36
Areca catechu, 20
Arenga, 12, 20, 21
artificial baits, 31
***Artocarpus* spp.**, 56
Asai Palm, 20
Asia, 12, 21, 24, 31, 41, 53, 111, 247, 302, 304
Asiko Grace Adala, 71
Assisted Pollination, 81
asthmatic, 20
atomella, 117
atripes, 116, 117, 135, 180
Australasian, 82, 247, 304
Australia, 247, 304, 333
Averhhoa carambola, 37, 69
avocado, 17, 18, 38
Avocado, 37, 38, 81
Azira Muhamad, 106
Bacang, 37, 59
Bactris gasipaes, 20
Baker, 114, 117, 122, 127, 131, 132, 134, 135, 136
bakeri, 122, 127, 128, 129
'Bali', 73
Balik Pulau, 113
Banana, 37, 39
banana tree trunks, 298
bananas, 12, 39
Bandjarmasin, 124
Bänziger, 31
Baphia nitida, 295, 297
barracks, 110
Barton, Canberra, 30
Basil, 149
Batang Padang, 131, 133, 134, 135, 136, 137, 138, 139, 30
Bats, 39
Batu Balai, 130, 131, 132, 133
Batu Caves, 130, 133
Batu Feringghi, 113
beach, 113, 295
bee biology, 257, 299
bee crowding, 298
beehives, 31, 243, 245
Beescape, 84, 256, 294, 296, 303
Beescape for Meliponines, 293, 303
belt, 117
Beluran, 139
benches, 294
Bernard DUPONT, 249
Beserah, 132
betel nut, 12
Betel Nut, 20
Bettotan, 131, 132, 133, 134, 136, 137, 138, 139
BfD Journal, 93
binghami, 137, 139, 142, 180, 253
Biochemistry, 148
Biosecurity issues, 235
Birds, 56, 113
biroi, 114, 305
Bisexual, 75
Bkt. Cherakah, 139
black, 117
  feet, 117
Black Bee Farm, 99, 100, 102
black head, 117
black netting, 104
black soldier flies, 104
Bleeding Hearts, 296
BMNH, 120, 122, 123, 126
Bogor, 244
Borassus flabellifer, 20
Borneo, 122, 124, 130, 131, 132, 133, 134, 136, 137, 138, 139, 143, 132, 133, 247, 252, 253, 254
Borneotrigona hobbyi, 142
Botanical gardens, 113
Botany, 301, 303
Bougainvillea, 299
Brazil, 121
Breadfruit, 37, 56
Brinjals, 37, 40
British Museum, 129, 31
broken corn, 295
Brunei, 33
BSF, 104
Bt. Timah, 131, 133, 137
Buds, 69
Bukit Cherakah, 131, 137
Bukit Jerneh, 138
Bukit Ketam, 36
Bukit Kutu, 130, 131, 132, 133, 134, 135, 136, 137, 138, 139, 131
Bukit Merah, 130
Bukit Panchor, 133, 135
Bukit Timah, 132
bunkers, 110
Burkino Faso, 75
Burma, 247
BURMA, 117
bushland, 19
Butia capitata, 20, 21
C. mollissima, 37, 43
cactus, 298
Calamansi, 24, 37
Calamondin, 24
calyces, 67
Camargo, 119, 122, 304
CAMBODIA, 128
Cameron, 9, 82, 117, 118, 132, 252, 253, 254
Cameron Highlands, 9, 131, 135, 140
canifrons, 131, 140, 180, 253
canola, 13
***Capsicum* spp.**, 37, 44
Carambola, 37, 69
Carica papaya, 37, 61
cashew, 17, 24, 42
Cashew, 37, 42, 81
cassava flour, 295
Castanea sativa, 37, 43
Caucasian chestnut, 43

Cave Nectar Bat, 50
cempedak, 149
Central America, 41, 49, 317
cerana, 66
cereals, 24, 316
Cervancia, 15, 301
Changi, 137
Chenomx, 108
Chestnut, 37, 43
Chiangmai University, 31, 33
Chief Minister of Melaka, 151
Chillies, 37, 44
Chintamani, 132, 133
chitin, 128
Chocolate, 41
Christmas season, 67
Ciku, 37, 74
Citrullus lanatus, 37, 78
Citrus, 17, 24, 37, 45, 65, 81, 135
clayey soil, 298
climbers, 293, 295
Cockerell, 114, 116, 117, 122, 126, 127, 128, 130, 140, 132, 252, 253, 254, 302
Cocoa, 37, 41
coconut, 13, 17, 22, 46, 84, 148, 149, 150, 151
Coconut Farm, 15
Coconut palms, 46
coconuts, 12, 21, 22
Cocos nucifera, 20, 21, 37, 46
Coffea Arabica, 47
coffee, 13, 17, 47, 133
Coffee, 37, 47, 81
collina, 136
collina, 136, 144, 139
collina, 126
colony resilience, 151
color, **124**
confusella, 143
container gardening, 77
Convolvulaceae, 296
Cooperatives, 31, 233
corn, 24, 57
cotton, 17, 44, 64, 111, 255
cotton jersey, 237
Craft Complex, 104
Crematogaster, 141
Crocker Range, 136, 137, 138
cross-pollinating, 46
Cross-pollination, 18, 68, 70, 78, 311
crushed peas, 295
***Cucumis* spp**, 37, 48
Cucurbits, 17, 37, 48, 81, 149
Cuphea, 296
darken
belt, 117
Date Palm, 20
Deborah Smith, 247
dessert, 100, 101
Diallo, 75
Dimocarpus longan, 37, 54
dioecious, 58, 73, 312
Dipterocarp, 139, 231, 248, 252, 254, 304
Dipterocarpaceae, 130, 247, 248
DNA Barcoding, 186, 188
doipaensis, 138, 180
Dover, 130, 131, 134, 136, 138
dragon fruit, 12, 49
Dragonfruit, 37, 49, 81
dresceri, 244
drifting swarms, 297
drooping clusters, 53
Durable boxes, 32
durian, 12, 13, 50, 113, 149
Durian Lai, 37, 50
Durian Perangin, 99
Durio zebethinus, 12
dusky, 114
Dusun Tua, 133
Dynastid, 68
East Malaysia, 247, 252
Ecology, 302, 304, 305
Eggplant, 37, 40
Elaeis guineensis, 20, 37, 60
elliptical, 53
endnote, 130
Engkabang, 249
Eonycteris spelaea, 50
erectile dysfunction, 100, 105
erythrogastra, 131, 252
Ethnoentomology, 94
Eurycoma longifolia, 99, 100, 108
Euterpe oleracea, 20
Expedition, 143, 31, 132
extrafloral nectar, 13, 298
Fac. de Agronomia e Medicina Veterinária, 71
Fajardo, 15
female flowers, 39, 43, 46, 48, 54, 55, 56, 58, 60, 66, 312, 321
Fenugreek, 100, 105
fertilization efficiency, 71
fig tree, 70
fimbriata, 132, 144, 180
flavibasis, 139
flesh percentage, 56
flies, 21, 41, 55, 59, 60, 257, 296, 319, 326, 327
flight patterns, 82, 333
floral canopy, 293
flowering weeds, 19, 84
fogging, 243, 256
fogging (anti-mosquito)., 242
foot, 117
foraging realm, 293
Forcipomyia midges, 41
Forcipomyiinae, 41
Forest Reserve, 138, 131
***Fragaria* spp.**, 71
FRANCE, 249
Francis Light, 114
Fraser's Hill, 121, 130, 132, 133, 135, 136, 138, 139, 36
fuliginous wings, 128
fulvomarginata, 122, 126
fuscibasis, 136, 142, 253
fusco, 117
fuscobalteata, 117, 143
Gaetan Biesuz, 101
gamat, 99, 100, 108
Garcinia mangostana, 37, 58
garden beds, 295
garlic, 99, 108
gas chromatography, 91
gazebos, 294
geissleri, 134, 253
Geniotrigona, 13, 82, 136, 180, 252
Geniotrigona lacteifasciata, 136
Geniotrigona thoracica, 13, 82, 180, 252
Genting Bidai, 130, 135, 136, 138
Genting Sempah, 131, 134, 136, 138
geopropolis, 298
George Town, 111, 116, 122, 126
Georgetown, 110
ginger, 99, 100, 105, 108
glossy paints, 245
Gn. Ledang, 30
Gombak, 137
Gombak Valley, 137, 138, 139
Gourds, 149
Grapefruit, 37, 65
Grapes, 37, 51
grapevines, 51
Great Farm, 13
Green, 67
greenhouse, 16, 33, 35, 40, 303, 304
Gribodo, 124, 125, 132, 253, 254
ground covers, 293, 295, 296
ground creepers, 293
guava, 17, 149
Guava, 13, 37, 52, 81
gum, 13
Gunong Jerai, 132, 134, 136
Gunong Keriang, 133
Gunong Kledang, 131
Gunong Tahan, 132, 133, 135, 136, 138
Gunung Ledang National Park, 126
Gunung Ledang Park, 31
H. M. Pendlebury, 130, 131, 132, 133, 134, 135, 136, 137, 138, 139

H. T. Pagden, 130, 131, 132, 133, 134, 135, 137, 138, 139
Hadi, 180
haematoptera, 143, 144, 254
haematoptera variety dulitae, 143, 144
Hainan, 247
halba, 100, 105
hand-pollinating, 40, 62, 73
haphazard transfer, 87
Hawaii, 22, 23
Heard, 302
heavy metal pollution, 245
hedgerows, 19
hedges, 293
Helianthus annuus, 37, 72
Heliconias, 295
Hemigraphis, 296
Henderson, 21, 130, 135
Hendra Ja, 82, 180
herbicides, 19, 242, 256
Hermaphroditic, 81
Heterotrigona, 13, 122, 127, 128, 131, 132, 136, 137, 180, 235, 252, 301, 302
Heterotrigona bakeri, 122, 127, 128, 132, 252
Heterotrigona erythrogastra, 131, 137
Heterotrigona itama, 13, 128, 137, 180, 252, 301
Hibiscus family, 50
Hibiscus sabdariffa, 37, 67
Highand Meliponicukture, 136
Highland, 130
highland agriculture, 11, 133
Hill Stream, 131, 134, 136, 138
hind tibia, 144
hives, 16, 31, 34, 35, 36, 38, 69, 84, 96, 99, 101, 133, 148, 236, 237, 239, 243, 334
Hj. Adly bin Zahari, 151
hobbyi, 144, 142, 254
hobbyist, 133
hobbyists, 232, 240, 241
homozygous, 46
honey
color, 124
honeybees, 19, 22, 43, 51, 63, 66, 310
honeydew, 298
Hot Spring Klang Gates, 131
hot sulphur spring, 133
Hrncir, 305
Hulu Selangor, 13
hummingbirds, 23, 63, 311, 319, 320, 325, 326
***Hylocereus* spp**, 37, 49
Hylocereus undatus, 49
Hymenoptera, 41, 66, 71, 120, 301, 302, 303, 304, 305
Ibrahim, 180
Ibrahim Shrub, 13
increase yield, 16
Increased Yield, 81
Indian sub-continent, 21
Indigenous knowledge, 106
Indo-China, 247
Indo-China 8 species, 247
Indo-Malaya, 247
INDONESIA, 117, 128
Indonesian archipelago, 21, 24
induced mortality, 300
inflorescences, 22, 46, 60, 73, 313
Inglis, 131, 132, 136, 139
Inoue, 71
insectivorous, 295
intermingling swarms, 297
Inter-Monsoon, 151
Introitus, 142
IPB, 244
Ipomoea Batatas, 296
iridescent
wings, 123
iridi, 123
iridipennis, 114, 123, 133, 143, 144, 132, 302, 305
itama, 84, 122, 128, 129, 130, 144, 235, 244
Jackfruit, 37, 56, 81
Jahudi Road, 110
Jalil, 9, 12, 22, 83, 84, 90, 31, 138, 142, 143, 144, 303, 332
Jamaica, 67
Jambos, 24, 37, 149
Japanese rose, 296
Jasmine, 149
Java, 247, 334
JAVA, 128
javanica, 138
Jeffry Iskandar Foo, 31, 180, 231
Jelly Palm, 20
Jerantut, 130, 131, 132, 133
Jewish, 110, 114
Jewish cemetery, 110, 111
Jitra, 131, 133, 135, 137
John Wyne James, 133
Johor, 23, 126
Johore, 132, 136, 139
Jor Camp, 130, 131, 133, 134, 135, 136, 137, 138, 139
K6dah, 133
Kabayau, 131, 133, 134, 137, 139
Kaffir Lime, 24
Kakutani, 71
Kanching, 132, 135, 137, 138
Kancing Hill, 36
Kangkung, 296
Karak, 133
Kedah, 104, 131, 132, 133, 134, 135, 136, 137
Kedah Peak, 132, 134, 136, 137, 131
Kelantan, 96, 235
Kelantan Biotech, 96
Kelapa Pandan, 99
Kelly, 133
Kelulut Honey, 148, 149, 151, 152
Kemburongoh, 142
Kenokok, 131, 134, 137, 139
Kenya, 71, 303
Kerling, 13
Kimura, 20
Kinabalu, 131, 133, 134, 137, 139, 133, 136, 139
King Bhumiphol, 249
Kloss, 130, 131, 132, 133, 134, 135, 136, 137, 138, 139
Kota Marudu, 139, 140
Kuala Kangsar, 133
Kuala Lipis, 130, 131, 132, 133
Kuala Lumpur, 131, 133, 135
Kuala Tahan, 130, 132, 133, 134, 135, 136, 137
Kuala Teku, 132, 133, 136, 139
Kuantan, 131, 133
Kubang Kerian, 96
Kudat, 134, 136, 137, 139
Kuini, 37, 59
Kundasang, 133, 136, 138, 143
kutejensis, 37, 50
L. aff *latipes*, 144
Labuan, 134
Labuk Highlands, 136, 137, 139
lacteifasciata, 82, 144, 252
laeviceps, 30, 244, 302, 334
landrace, 299
Langkawi, 94, 95, 96, 100, 101, 103, 104, 106, 107, 108, 131, 133, 134, 137
Langkawi Department of Agriculture, 104
Langsat, 37, 53
Lansa, 53
Lansium domesticum, 37, 53
Lanzon, 53
Lanzones, 53
Larut Hills, 130, 132, 133, 134, 137, 138, 142
latebalteata, 139, 143, 138, 254
latex, 13
Latin, 117, 123, 124, 126
lawn, 295
lead-based, 245, 257
leaviceps, 143
Ledang, 126
legumes, 17, 81
lemon, 13, 45, 99, 100, 105, 108

Lepidoptera, 41
Lepidotrigona, 13, 122, 126, 31, 132, 136, 138, 144, 180, 253, 254
Lepidotrigona terminata, 13, 122, 126, 132, 253
Liangtelan, 124
Lime, 24, 37
Lisotrigona, 33, 253
Litchi chinensis, 37, 55
lizards, 296
Loam, 298
Longans, 37, 54, 81
Lophotrigona, 136, 140, 180, 253
Lowe, 130, 133, 135, 136
LOWLAND SLOPES, 143
Lubok Tamang, 130, 135
lychee, 17, 55
Lychee, 37, 55, 81
Maeta, 71
maintenance regime, 16, 35
Maize, 24, 37, 57
Malacca, 116, 124, 126, 235
Malacca state airport, 148
malar space, 129
Malay, 110, 126
Malay Peninsula, 247
Malaya, 121, 130, 136, 231, 247
Malaysia, 126
Malaysia Genome Institute, 108, 31, 169
MALAYSIA- PENINSULA, 117, 128
male congregation, 298
male flowers, 38, 39, 46, 48, 55, 61, 66, 312
Malinsau, 140, 142, 144
Malvaceae, 50, 67
Mandai, 131, 135
mango, 13, 17, 59, 149, 305, 318
Mango, 37, 59, 81
Mango farm, 15
mangosteen, 58, 149
Mangosteen, 37, 58, 81
Manikara zapota, 37, 74
MARDI, 13, 14, 233, 235, 303
Markisa, 37, 62
Mass Spectrometer, 107
Maxwell's Hill, 130, 133
Megachilidae, 99
Melaka, 148, 151, 152
melanocephala, 143, 144, 254
melanochepala, 142
melanoleuca, 138, 139, 180
Melastoma, 130, 131
melina, 124, 144
Melipom basimaculata, 141, 132
Melipona, 305, 317
Meliponary, 84, 101, 293, 295, 296, 297, 334
Meliponiculture technique, 87
Meliponine repository, 170
Meliponini, 71, 202, 301, 302, 304, 305
mellifera, 75
melon, 13, 14, 17, 78
***Mengifera* spp.**, 37, 59
Merlimau, 148, 149, 150, 151, 152
Merlimau Kelulut Honey Farm, 148, 152
metabolites, 106, 107, 108
MGI, 106, 31, 169, 170
Michener, 30, 254, 303
midge, 41
Millennium Ecosystem Assessment, 13
Miller, 130, 131, 132, 133, 139
Mimosa, 296
Min Camp, 96
minangkabau, 244, 254
Mitchell, 22
Mohd Nahizam b Nosrudin, 104
Mohd Razif Mamat, 106
Momose, 303
monkeys, 241
monoecious, 21, 46, 48, 57, 73, 318
Monofloral, 150
Monsoon, 151
moorei, 144, 141, 244, 253
morning glory, 296
morphologically, 316
Morton J, 12, 23
Moss, 136
Moss Forest, 144
Mount Kinabalu, 133
MSNG, 124
Mt Dulit, 143
Mt Ophir, 30
Mt. Ophir, 116, 136, 30, 31
Muar, 126
Mucilage, 298
mucus, 298
Muhammad Fahdli, 148
Mukim Bohor, 104
multi-floral, 150
***Musa* spp**, 37, 39
My GAP, 9, 31
National Institutes of Biotechnology Malaysia, 188
Natural History Museum, London, 30
nectar, 13, 21, 22, 40, 43, 50, 56, 66, 70, 74, 81, 92, 149, 150, 151, 234, 295, 298, 299, 312, 318, 319, 320, 322, 326, 327
nectaries, 21
Negri Sembilan, 139
Nephelium lappaceum, 37, 66
New South Wales, 247
Niah, 249
Nicola Bradbear, 93
night bloomers, 49
nitidiventris, 138, 31, 253
NMR, 106, 107, 108
nocturnal flowering, 12
nonverbal communication, 295
Nor Azfa Johari, 106
Norhasnida Zawawi1, 200
Norowi, 13, 303
Nuclear Magnetic Resonance, 107
Nurul Ainaa Farhanah Mat Ramlan, 200
Odirce Teixeira Antunes, 71
Odontotrigona, 254
Odontotrigona cf *haematoptera*, 143
Oecophylla smaragdina, 75
oil palm, 20, 60
Oil Palm, 20, 60
okra, 17, 298
One-Stop Information Centre, 233
Ophir
Mount, 126
opium poppy, 17
orchard, 38, 69, 305
organic farming, 17, 36
OUMNH, 116, 122
pagdeni, 120, 142, 254, 305
pagdeniformis, 120
Pahang, 130, 131, 132, 133, 134, 135, 136, 137, 138, 139
paints, 245
palm crops, 12
palm spadices, 293
Palmyra fruit, 20
Papaya, 37, 61
par course, 294
Passiflora edulis, 37, 62
passion fruit, 17, 62
Passion Fruit, 37, 62, 81
Pawpaw, 37, 61
Peach Palm, 20
pemangensis, 114
Penampang, 141
PENANG, 128
Penang Free school, 112
Penang Hill, 113
Penang Island, 110, 114, 117, 122, 127
Penati, 124
peninsularis, 138
Peperomia pellucida, 99, 100, 105, 108
Pepper Elder, 99, 108
Perak, 124, 130, 131, 132, 133, 134, 135, 136, 137, 138, 139
perennial plants, 295
Perlis, 138
Persea americana, 37, 38

Pesticide Regime, 33
pesticides, 16, 34, 35, 242, 256
phloem, 13
Phoenix dactylifera, 20
physiognomy, 300
physogastric, 87
Pili farm, 15
Pinang Palm, 114
Pineapple, 12, 23, 37, 63
pineapple plantation, 23
Pinousuk Peak, 136
pistillate, 21, 22, 61, 314, 321, 327, 329
Pitaya, 37, 49
Pittaya, 12
Plant pollination, 13
plant science, 148
playground, 294
pliable tubes, 112
pollen, 13, 18, 20, 21, 22, 24, 38, 40, 42, 43, 44, 48, 51, 55, 57, 59, 60, 61, 62, 63, 64, 65, 66, 68, 70, 71, 72, 73, 74, 78, 81, 93, 144, 234, 298, 309, 310, 311, 312, 313, 314, 315, 316, 317, 318, 320, 321, 322, 323, 324, 325, 326, 327, 328, 329, 330
pollen grain, 24, 316, 318, 322, 323, 327, 328
pollinating agents, 11, 13, 311
Pollination Initiative, 16, 35, 301
pollinizer, 18, 326
Polyvinyl Chloride, 91, 305
Pomegranate, 37, 64, 81
Pomelo, 37, 65
Pomelos, 24
Population Ecology, 71
Poring, 136, 139
Port Dickson,, 139
Portable Cabins, 36
Portuguese, 113
Portulacas, 296
powdered nuts, 295
predators in flight, 295
primary agents, 81
principal pollinators, 23
profit increase, 17, 35
propolis, 9, 12, 20, 82, 83, 90, 91, 92, 104, 142, 149, 257, 332, 333
Psidium guajava, 37, 52
Pulau Jarak, 134
Pulau Perhentian, 134
Punica granatum, 37, 64
pure black, 114
Purseglove, 21
Purslane, 296
PVC, 91, 92, 305
Rabbi, 110
racemes, 295
rachillae, 21
Radzi Alias, 95
Raimy Ché-Ross, 30
Ramalingam, 74
ramblers, 293
Rambutan, 37, 66
Rambutans, 24
Ranau, 133, 136, 141, 142, 143, 144
Rasmussen, 114, 117, 120, 121, 123, 124, 126, 30, 252, 302, 304
Ravenala Palms, 295
Ravenella, 297
red ant, 75
reepeni, 143, 254
Relative humidity, 295
rental rates, 14
Rental rates, 34
resin, 13, 53, 130, 139, 149, 151, 231, 233, 234, 245, 247, 248, 299
Resin producing Trees, 248
rest huts, 294
Rhoeo, 296
roadside verges, 19
Robusta coffee, 47
Ronaldo Zucchi, 121
Roselle, 37, 67, 81
Roubik, 9, 13, 31, 303, 304, 305, 310, 311, 314, 315, 316, 318, 321, 322, 323, 325, 327, 328, 329, 330
rough verges, 19
rubber, 19, 245, 257
rubber plantation, 19
rubbish bins, 294
rufibasalis, 136, 144, 254
Sabah, 36, 133, 136, 302
Sakagami, 120, 121, 132, 254, 304, 305
Salacca, 20, 21, 37, 73
***Salacca* spp.**, 73
Salacca zalacca, 20
Salak, 20, 37, 73
salts, 31
Samanea saman, 104
Samawang, 131, 132, 134, 136, 137
Sambamurty, 74
Sandakan, 131, 132, 133, 134, 136, 137, 138, 139
sap, 13, 33, 37, 70, 299
Saplings, 293
Sapodilla, 37, 74
Sarawak, 23, 122, 143, 132, 248, 249
sarawakensis, 144, 143, 254
Sawo, 37, 74
scabbard, 142
scarab beetles, 68
scarce, 145
Schwarz, 117, 120, 128, 130, 131, 132, 133, 134, 135, 136, 137, 138, 139, 143, 31, 132, 247, 253, 254, 305
sea cucumber, 99
seeds and oil, 72
SEHU, 121
Seimund, 130, 131, 132, 133, 134, 135, 137, 138, 139
Selangor, 130, 131, 132, 133, 134, 135, 136, 137, 138, 139, 36, 131, 200, 231
self-fruitful, 39, 45, 52, 64
self-incompatible, 74, 75, 328
self-pollinating, 40, 44, 46, 47, 57, 65, 67, 68, 72, 77, 313, 328
self-pollination, 12, 22, 38, 45, 49, 56, 57, 59, 61, 311, 315
Selitar, 137
seminars, 15
Sepang, 134
Sg. Jang, 13
Shade tolerant, 296
Shamsul Bahri, 141, 143
Sholdt, 22
Shorea, 248, 249
Shorea roxburghi, 249
Shrubbery, 293
Shuib, 22, 180, 303
Siam, 117, 130
Singapore, 20, 129, 131, 132, 133, 135, 137, 30, 303
SINGAPORE, 128
Singapore Botanic Gardens, 133
sirindhornae, 254
Siti Salmah Idrus, 244
sluggish, 100, 105
small beetles, 68
Smith, 82, 116, 117, 119, 122, 123, 126, 142, 31, 132, 252, 253, 302, 305
Smuggling logs, 236
Snake skin fruit, 20
Snakeskin fruits, 12
Solanum lycopersicum, 37, 76
Solanum melongena, 37, 40
Solenopsis, 66
sorghum, 24, 57
Soursop, 37, 68
South America, 21, 41
South East Asia, 21, 247
Spectra, 108
spiders, 296
splitting, 87
sports hobbies, 295
squamosa, 37, 68
Sri Lanka, 21, 304
St. Regis, 100, 101
Staminate, 12, 21

star fruit, 13
Starfruit, 37, 69
Statistics, 20
Steppingstones, 294
strata levels, 293
Strawberry, 71
Strelitzias, 295
Structural shrubs, 293
Sudanian Region, 75
sugar, 31, 107, 204, 319
Sugud river, 136, 137, 139, 140
Sumatra, 247, 305
SUMATRA, 117, 128
Sundatrigona, 136, 141, 244, 253
sunflower, 17, 72, 301
Sunflower, 37, 72, 81
Sunflower seed, 72
Sungai Chongka', 139
Sungai Liwagu, 136
Sungai Palong, 132, 136, 139
Sungai Pinang Hill, 30
Sungai Rengi, 148
Sungai Sugud, 136, 142
Sungei Tahan, 130, 132, 133, 134, 136, 138
swales, 295
Sweet potato, 296
Sweet Shorea, 249
swelling, 100, 105
Syzygium, 149
***Syzygium* spp.**, 37, 70
Tahan River, 133
Taiping Hill, 130, 133, 134
Taiwan, 247
Takehiko, 71
Tamarind, 37, 75
Tamarindus indica, 37, 75
Tambunan, 133, 136
Tamiji, 71
Tamparuli, 136, 141
Tanah Rata,, 131, 134, 135
Tangarine, 24
Tangkak, 31
Tanjong Mas, 134
Tegula, 114
Teluk Yu, 104
Telupid, 136
Tembeling, 132
template, 14, 32, 33
TENASSERIM, 117
Tenompok Pass, 131
terminata, 88, 139, 143
termite, 83
termites, 257
Tetragona, 120
Tetragonilla, 13, 116, 117, 126, 132, 136, 139, 253, 254
Tetragonilla atripes, 13, 116, 117, 132, 253
Tetragonilla cf *rufibasalis*, 143
Tetragonilla collina, 126, 132, 253
Tetragonula, 13, 15, 114, 117, 118, 119, 120, 121, 123, 124, 125, 140, 30, 33, 132, 136, 142, 143, 180, 244, 253, 254, 302, 304, 305, 334
Tetragonula biroi, 15, 302
Tetragonula fuscobalteata, 117, 118, 132, 180, 253
Tetragonula geussleri, 140, 132
Tetragonula iridipennus, 30
Tetragonula laeviceps, 13, 119, 132, 180, 253, 302
Tetragonula melina, 124, 125, 132, 253
Tetragonula pagdeniformis, 120, 36, 132, 254
Tetragonula penangensis, 114, 132, 254
Tetragonula zucchii, 121, 132, 254
Tetrigona, 13, 122, 142, 30, 33, 132, 136, 139, 180, 241, 253
Tetrigona apicalis, 13, 122, 142, 132, 180, 253
Tetrigona binghami, 36
Tetrigona peninsularis, 13, 30, 253
Tezuka, 71
Tg Malim, 30
Thailand, 50, 66, 235, 236, 247, 249, 254, 301, 302, 305
THAILAND, 117, 128, 301
The Gap, 131, 132
Theobroma cacao, 37, 41
Theodore Dru Allison Cockerell, 114
thoracic, 84
thoracica, 82, 84, 87, 88, 90, 133, 144, 142
thorax, 117, 299, 322
Tioman Island, 132
toilets, 294
tomato plants, 77
tomatoes, 13, 77, 305, 310
Tomatoes, 37, 76, 77
tombstones, 110
Tongkat Ali, 99, 100, 105, 108
Topping, 15
Topsoil, 295
Toshiyuki, 71
toxicity, 19
TRAC, 96
Training Retreat, 96
transit time, 87
treehoppers, 298
trehalulose, 204
Trigona, 114, 116, 117, 120, 121, 122, 123, 124, 126, 127, 128, 130, 143, 144, 132, 301, 302, 304, 305, 317
Trigona atomella, 117
Trigona minangkabau, 71
trochanterica, 138, 254
Tropical Fruits, 23
Trus Madi, 139
Trus Madi Range, 136, 139
Tuaran, 36
U. S. National Museum, 129
Ulu Gombak, 131, 132, 135, 137, 131
Ulu Klang, 133
Ulu Langat, 135, 139
Ulu Sangai Triang, 136
Ulu Sungai Triang, 131
UMT, 143
unisexual flowers, 21
Universiti Putra Malaysia, 200
University of Nairobi, 71
University of Sao Paulo, 121
USNM, 117, 122, 127
valdezi, 134
ventralis, 144, 138, 180, 253
Vietnamese giant, 49
Vinca, 296
Visit Melaka Year 2019, 151
Vitis vinifera, 37, 51
volume of hive boxes, 244
walkways, 294
Wallacetrigona. incisa, 244
wasps, 21, 55, 59
waste fluids, 295
water body, 294
Water spinach, 296
Watermelons, 37, 78
wax, 82, 90, 92, 149, 257, 317, 333
weather conditions, 87, 151, 297, 334
West Malaysia, 247
White, 83
wild animal predators, 241
wild bananas, 39
wild boars, 241
wild pollinators, 19
wind pollination, 22, 55, 60, 70, 309, 321
windbreakers, 295
Wray's Camp, 132, 133, 135, 136, 138
Wray's Hill, 138
Yasuo, 71
Yoshikawa, 116, 122, 126
Yunnan, 247
Zea mays, 24, 37, 57
Zephyranthes, 296
Zucchi
Prof. Dr. Ronaldo, 121
zucchii, 121

www.ingramcontent.com/pod-product-compliance
Lightning Source LLC
LaVergne TN
LVHW071200210926
848482LV00014B/697